AF567153

Structure Energetics and Reactivity in Chemistry Series (SEARCH Series)

Series Editors

JOEL F. LIEBMAN
Department of Chemistry and Biochemistry
University of Maryland Baltimore County
Baltimore, MD 21228

ARTHUR GREENBERG
Department of Chemistry
University of North Carolina at Charlotte
Charlotte, NC 28223

The volumes in this series are comprised of state-of-the-art reviews, explicitly pedagogical in nature, in which specific topics are treated in depth. The series acronym SEARCH reflects the interplay between *S*tructure, *E*nergy *A*nd *R*eactivity in *CH*emistry and how these are also manifested in physical properties and biological activities.

Other titles in the Series

Volume II
Active Oxygen in Chemistry
Edited by Christopher S. Foote, Joan Selverstone Valentine, Arthur Greenberg, and Joel F. Liebman

Volume III
Active Oxygen in Biochemistry
Edited by Joan Selverstone Valentine, Christopher S. Foote, Arthur Greenberg, and Joel F. Liebman

Editorial Advisory Board

MESOMOLECULES

FROM MOLECULES TO MATERIALS

SEARCH Series, Volume 1

EDITORS:

G. DAVID MENDENHALL
Michigan Technological University

ARTHUR GREENBERG
University of North Carolina at Charlotte

JOEL F. LIEBMAN
University of Maryland Baltimore County

CHAPMAN & HALL

 An International Thomson Publishing Company

New York • Albany • Bonn • Boston • Cincinnati • Detriot • London • Madrid • Melbourne • Mexico City
Pacific Grove • Paris • San Francisco • Singapore • Tokyo • Toronto • Washington

Cover design: Trudi Gershenov

For more information, contact:

Chapman & Hall
One Penn Plaza
New York, NY 10119

International Thomson Publishing
Berkshire House 168-173
High Holborn
London WC1V 7AA
England

Thomas Nelson Australia
102 Dodds Street
South Melbourne, 3205
Victoria, Australia

Nelson Canada
1120 Birchmount Road
Scarborough, Ontario
Canada M1K 5G4

International Thomson Editores
Campos Eliseos 385, Piso 7
Col. Polanco
11560 Mexico D.F. Mexico

International Thomson Publishing
Gmbh
Konigwinterer Strasse 418
53228 Bonn
Germany

International Thomson Publishing Asia
221 Henderson Road #05-10
Henderson Building
Singapore 0315

International Thomson Publishing—Japan
Hirakawacho-cho Kyowa Building, 3F
1-2-1 Hirakawacho-cho
Chiyoda-ku, 102 Tokyo
Japan

1 2 3 4 5 6 7 8 9 10 XXX 01 00 99 98 97 96 95

Library of Congress Cataloging-in-Publication Data

Mesomolecules : from molecules to materials / editors G. David Mendenhall, Arthur Greenberg, and Joel F. Liebman.
p. cm. — (Structure energetics and reactivity in chemistry ; v. 1)
Includes bibliographical references and index.
DOI 10.1007/978-94-011-0005-2
1. Macromolecules. I. Mendenhall, G. David.
II. Greenberg, Arthur. III. Liebman, Joel F. IV. Series.
QD381.M47 1995
547.7—dc20 94-13543
CIP

British Library Cataloguing in Publication Data available

Please send your order for this or any other Chapman & Hall book to **Chapman & Hall, 29 West 35th Street, New York, NY 10001, Attn: Customer Service Department.** You may also call our Order Department at 1-212-244-3336 or fax your purchase order to 1-800-248-4724.

For a complete listing of Chapman & Hall's titles, send your request to **Chapman & Hall, Dept. BC, One Penn Plaza, New York, NY 10119.**
www.springer.com/mycopy

Contents

Preface

The title of this volume implies a progression of sorts from species of molecular size to a product described on the basis of continuum properties. The difference in approach from the standpoint of molecular behavior, on the one hand—more the forte of chemists—and from the standpoint of large-scale properties, on the other—more the province of chemical engineers and materials scientists—represents a severe cultural divide, but one with much potential for creative input from both sides.

Chapter 1 of this volume attempts a broad survey of trends toward the synthesis of large, well-defined molecular systems with interesting physical, chemical, or material properties. Review articles with more detailed treatments are emphasized. In Chapter 2, Newkome and Moorefield summarize work on synthesis of "cascade" molecules. Next, Denti, Campagna, and Balzani describe the synthesis of assemblies with connected metal-containing chromophore units which transmit electrons or electronic energy in defined ways. In Chapter 4 Wuest describes the construction of hydrogen-bonded organic networks, and in Chapter 5 Michl defines a molecular-level construction set. Finally, Jaszczak points out how nature's attempts over geological time spans are emulated by recent human synthetic activity in the fullerene arena, through the appearance of various morphologies of natural graphite.

The book concludes with a method for describing fractal-like molecules, and an index based on the method for appropriate compounds described in the text.

DAVID MENDENHALL
January 1994

Series Preface

The purpose of this series is the presentation of the most significant research areas in organic chemistry from the perspective of the interplay and inseparability of structure, energetics, and reactivity. Each volume will be modeled as a text for a one-semester graduate course and will thus provide groundwork, coherence, and reasonable completeness. In this context, we have made the editorial decision to defer to the authors the choice of the desired blend of theory and experiment, rigor and intuition, practice and perception. However, we asked them to engage in the spirit of this venture and to explain to the reader the basis of their understanding and not just the highlights of their findings. For each volume, and each chapter therein, we have aimed for both a review and a tutorial of a major research area. Each volume will have a single theme, unified by the common threads of structure, energy, and reactivity for the understanding of chemical phenomena.

Structure, energetics, and *reactivity* are three of the most fundamental, ubiquitous, and therefore seminal concepts in organic chemistry. The concept of structure arises as soon as even two atoms are said to be bonded, since it is there that the concept of bond length and interatomic separation begins. Three-atom molecules already introduce bond angle into our functioning vocabulary, while four atoms are needed for the introduction of the terms *planarity, nonplanarity,* and *dihedral angles.* Of course, most organic chemists are interested in molecules of more than four atoms, so new shapes (tetrahedra, cubes, dodecahedra, prisms, and numerous exotic polyhedra) and new degrees of complexity arise. These new molecular shapes, in turn, function as templates for the next molecular generation. Still, the basic assumption remains: molecular structure determines energy and reactivity, and even though Van't Hoff, LeBel, and Sachse explained chemical reality with palpable molecular models over a century ago, we still do much the same thing on the screens of personal computers.

The concept of energetics arises in chemistry as soon as there is a proton and an electron and remains with us throughout our discipline. There are the fundamental, experimentally measurable quantities of bond energies, proton affinities, ionization potentials, pK_a values, and heats of formation. There are the derived quantities such as strain and resonance energies, acidity, and basicity. There are also the widely used, generally understood, and rather amorphous concepts such as *delocalization, conjugation,* and *aromaticity.* Indeed, the shape, conformation, and therefore function of a protein are determined by a balance of energetics contributions—resonance in the peptide linkage, hydrogen bonding, hindered rotation of a disulfide, Van der Waals forces, steric repulsion, Coulombic interactions and salt bridges, and solvent interactions.

The concept of reactivity inseparably combines structure and energetics and introduces more concepts and words: *stereospecificity, intramolecularity, nucleophilicity, catalysis, entropy of activation, steric hindrance, polarizability, hard and soft acids and bases, Hammett/Taft parameters.* Reactivity is a more difficult concept than structure and energetics. One must specify reaction conditions and usually accompanying reagents, since reactivity generally refers to two species or at least two seemingly disjoint parts of the same molecule.

Structure, energetics, and reactivity in chemistry have been probed by a plethora of experimental and theoretical methods. These tools have different degrees of accuracy and applicability, and consensus is rare as to when our understanding is deemed adequate. Indeed, diverse approaches—heats of hydrogenation and Hartree-Fock calculations, line intensities and LD_{50} values, ease of substitution and of sublimation, coupling constants and color—all contribute to the special blend of rigor and intuition that characterizes modern organic chemistry.

As people, and not just as scientists and editors, we wish to acknowledge the unity of the intellect and the emotions. We are grateful for the love and support of our families and for the inspiration, agitation, and stimulation from our mentors, colleagues, and students, and so we dedicate these volumes *"To Research and to Reason, To Family and to Friendship."*

JOEL F. LIEBMAN
Baltimore, Maryland

ARTHUR GREENBERG
Charlotte, North Carolina

Contributors

Vincenzo Balzani
Dipartimento di Chimica "G. Ciamician" dell'Università
40126 Bologna, Italy

Sebastiano Campagna
Dipartimento di Chimica Inorganica e Struttura Molecolare dell'Università
98166 Vill. S. Agata, Messina, Italy

Gianfranco Denti
Laboratorio di Chimica Inorganica
Istituto di Chimica Agraria dell'Università
56124 Pisa, Italy

John A. Jaszczak
Department of Physics and the Seaman Mineral Museum
Michigan Technological University
Houghton, MI 49931

G. David Mendenhall
Department of Chemistry
Michigan Technological University
Houghton, MI 49931

Josef Michl
Department of Chemistry and Biochemistry
University of Colorado
Boulder, CO 80309

Charles N. Moorefield
Center for Molecular Design and Recognition
Department of Chemistry
University of South Florida
Tampa, FL 33620

George R. Newkome
Center for Molecular Design and Recognition
Department of Chemistry
University of South Florida, Tampa, FL 33620

James D. Wuest
Département de Chimie
Université de Montréal
Montréal, Québec, H3C 3J7 Canada

MESOMOLECULES

FROM MOLECULES TO MATERIALS

SEARCH Series, Volume 1

1

From Molecules to Materials

G. David Mendenhall

THE RISE OF NANOSCALE ASSEMBLIES

When Is a Molecule a Material?

"One day the chemists will meet the physicists," a quote whose source I cannot remember, refers to the tendency of chemists to deal with ever-larger molecules, while physicists are engaged in the study of progressively smaller molecular aggregates. One may ask if the dimension of *size* has a direct relation to our theme, and whether there is a certain *molecular* size that signals the transition between molecule and material.

In fact, this is a common question that has been answered in many different ways. The answer depends on the material property in question. For solutions of polymers, for instance, the transition between small molecule and polymer is identified precisely as the point where chain entanglement results in a change in the relation between viscosity and molecular weight (Figure 1-1). Above a *critical molecular weight,* the viscosity increases as $MW^{3.4}$, compared with $MW^{1.0}$ below this critical value. The critical value depends somewhat on the nature of the polymer and the solvent, but in general, a critical molecular weight of around 10^4 is observed. For a given polymer and solvent, there is a corresponding *critical degree of polymerization* where the transition occurs that can be identified as an exact integer value.[1]

Similar distinct changes in slope are observed in plots of the glass transition temperature of various polymers vs. molecular weight,[2] each showing three linear regions of decreasing slope (Figure 1-2). Thus a

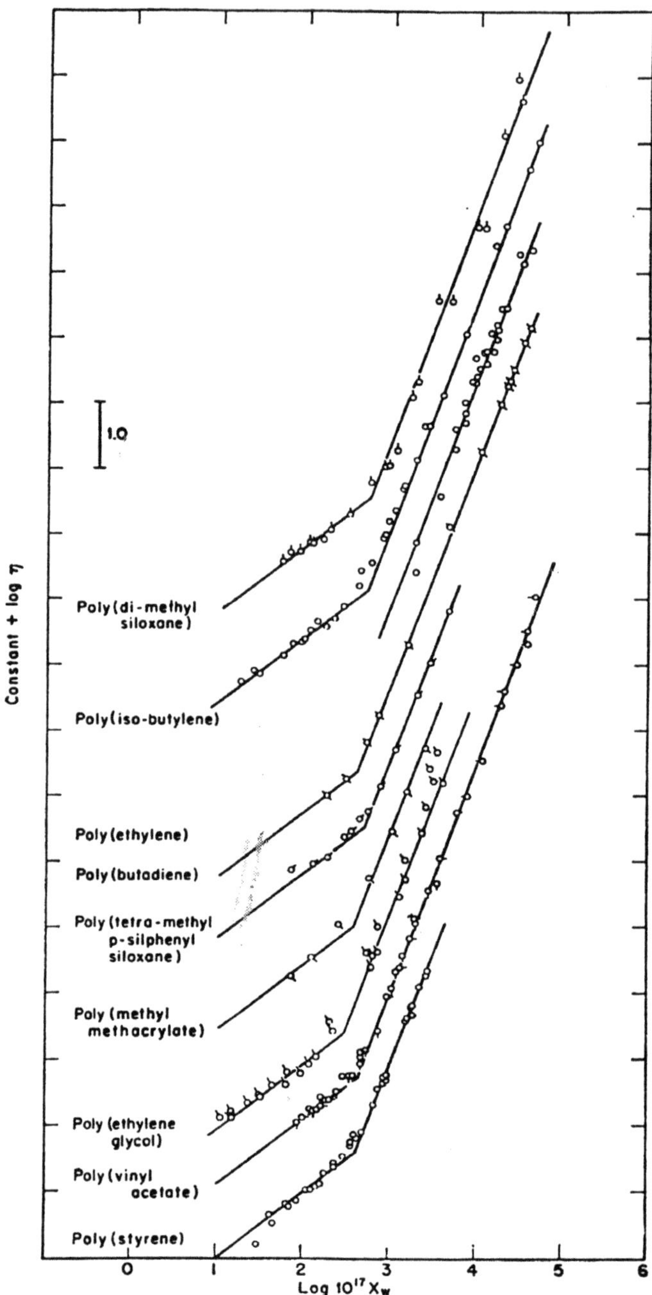

FIGURE 1-1 Plot of viscosity vs. molecular weight for a series of polymers. [From G.C. Berry and T.G. Fox, *Adv. Polymer Sci.* 5, 261 (1968)].

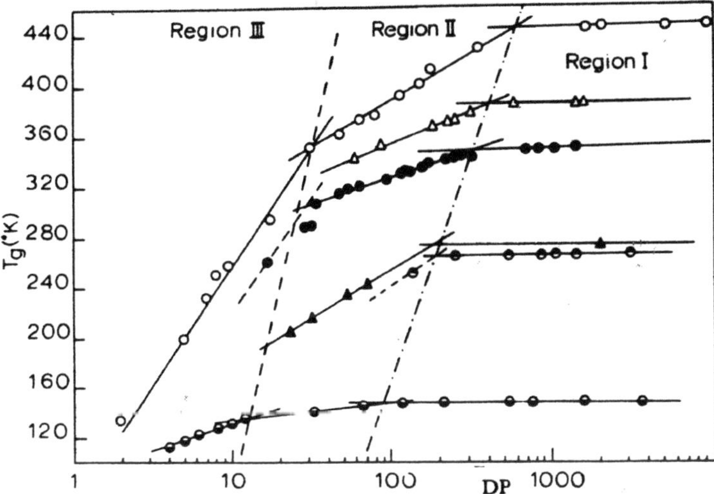

FIGURE 1-2 Plot of T_g vs. molecular weight for a series of polymers. [Reprinted from J.M.G. Cowie, *Eur. Polym. J.*, **11**, 297 (1975) with kind permission from Elsevier Science Ltd, The Boulevard, Langton Lane, Kidlington OX5 1GB, UK]

molecule may be said to become a material when increases in its size cause changes in properties corresponding to the large-molecule limit—which in this case is no change at all.

For metal atoms, the beginning of the transition from extremely reactive atomic behavior to the milder chemical behavior of a bulk metal sample has been associated with clusters large enough so that the atoms in the center have only other metal atoms as neighbors.[3]

The question of when a molecule becomes a material can be answered in a radically different but equally important way by reference to the *temperature.* Most students have been entertained at some point by lecture demonstrations of the dramatic transition of rubber tubing or bananas into ceramic materials after immersion in liquid nitrogen. The transition of mobile liquids to a disordered, glassy state upon cooling is a feature of organic polymers, Pyrex, and metallic palladium. The changes in non-metallic material properties that accompany the change in temperature are described by the (WLF) equation:

$$\log a_T = \frac{-8.86\,(T - T_s)}{101.6 + T - T_s}$$

Here a_T is the ratio of all mechanical and electrical relaxation times for temperatures T within about 50 °C of a reference temperature T_s

unique for each substance. This master equation was published by Williams[4] in 1955 as a single smooth curve containing points from 18 polymers and polymer-solvent combinations, and was elaborated that year by Williams, Landel, and Ferry.[5] The equation predicts that the temperature dependence of the ratio of material properties (dynamic mechanical, electrical relaxation, etc.) follows a universal pattern defined by a reference temperature unique for each material. The equation in its original presentation could accommodate *n*-propanol (m.p. −127 °C), rubbery polymers, and silicate glasses, even though at room temperature these substances would be subdivided into one or the other group according to our notions of chemicals vs. bulk materials. In fact, the master-curve relationship illustrated in part in Figure 1-3 for polymers and inorganic glasses represents something profound when one notes that the principal intermolecular forces are ionic attractions in an inorganic glass, hydrogen-bonding in *n*-propanol, and London and van der Waals forces in the hydrocarbon polymer. Clearly, important aspects of the underlying behavior are independent of the molecular detail, which affects only the reference temperature, and are sensitive to the magnitude of the intermolecular forces with respect to RT. In summary, a molecule also becomes a material simply by lowering its temperature.

The result illustrated in Figure 1-3 demonstrates why traditional practitioners of material science and rheologists are frequently able to remain immersed in mathematical representations, aloof from the molecular concerns that intrigue chemists, even exasperating chemists by their lack of interest in and inability to describe the chemical formulas of compounds with which they are working. At the same time, material scientists feel perfectly at home with asphalt and bread dough, whose molecular complexity most chemists, possibly analytical ones excepted, might find unappetizing.

The splendid example of living systems, and their ability to muster the molecular detail to organize and distinguish among myriad biological molecules and polymers, cannot be explained away with a master equation. The rich detail of natural systems is an inspiration for what soon will be achieved *in vitro.*

Trends Toward Supermolecular Synthesis

Perusal of the recent literature in connection with the theme of this book revealed, in addition to much traditional synthesis of organic and polymeric compounds, a pattern of simple paths to ever larger structures. The classical synthesis of chlorophyll by Woodward and 18 coworkers, for instance, represented a combined effort of many years of labor[6] (Figure 1-4).

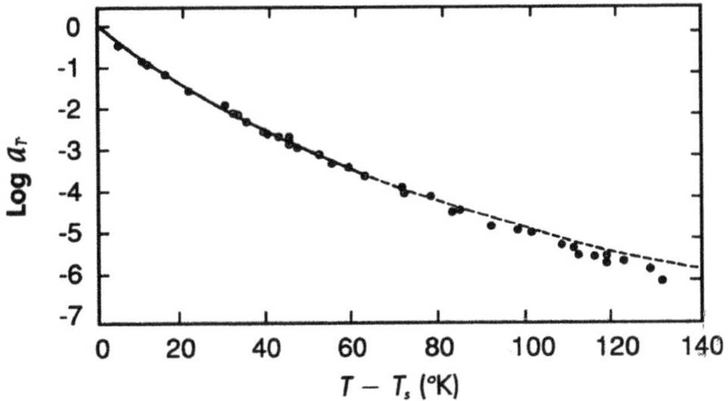

FIGURE 1-3 Master plot of reduced viscoelastic properties against reduced temperature according to the WLF equation.[5,6] Data points are for silicate and boron trioxide glasses, and the superposed line is derived from data on 17 organic polymers.[5,6]

However, examination of the structure of chlorophyll at once reveals symmetrical features, the modified pyrroles about the metal atom, while the peripheral groups facilitate the biological synthesis or the role of the compound in an energy-generating chloroplast. A representative of the porphyrin system underlying the symmetrical structure of chlorophyll can be synthesized quite simply as a short undergraduate experiment from pyrrole and benzaldehyde under acidic catalysis in a one-pot reaction.[7] A lucky investigator can achieve the facile construction of complicated molecular systems, as shown in two examples in Figure 1-5.

The syntheses of "picnic basket" porphyrins[8]—that is, porphyrins with functionalized belts spanning an iron atom in the center as mimics for cytochrome P-450[9,10]—and membrane-spanning metalloporphyrins[11] were facilitated by the comparatively easy construction of the underlying

FIGURE 1-4 Structure of chlorophyll a.

$CH{=}CH_2$
Me
Me
Et
N
N
Mg
N
N
Me
H
Me
$PhytylOOCCH_2CH_2$
H
MeOOC
O

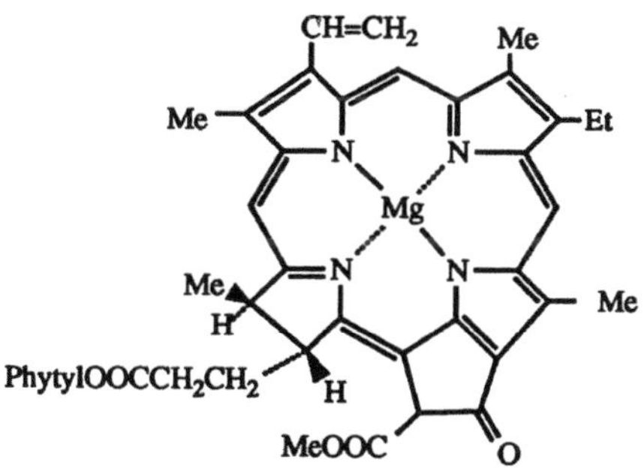

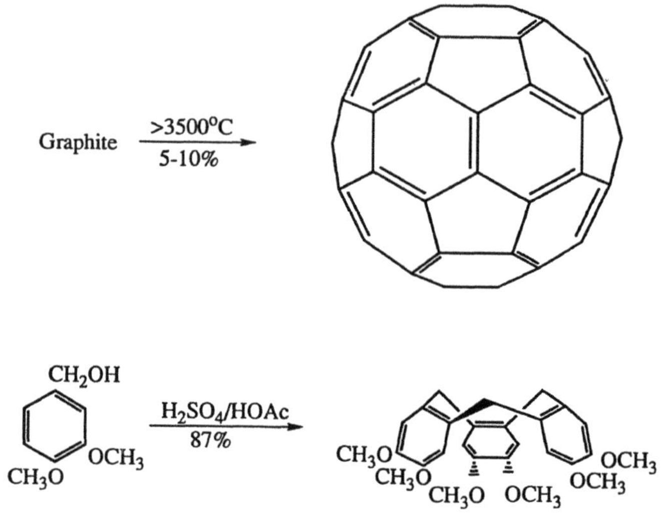

Graphite $\xrightarrow[5\text{-}10\%]{>3500^{\circ}C}$

CH_2OH, OCH_3, CH_3O $\xrightarrow[87\%]{H_2SO_4/HOAc}$ CH_3O, CH_3O, CH_3O, OCH_3, OCH_3, OCH_3

FIGURE 1-5 Two complicated molecules with easy syntheses.

heterocyclic ring system. Similarly, the combination of phenanthroline-like ligands into molecular assemblies (Chapter 3, this volume) is a compromise between these extremes, combining elegance and facility in a synthetic approach to molecules that eventually may function the way chlorophyll does to harvest light energy.[12]

The syntheses of arborols (Chapter 2, this volume), dendrimers,[13] fullerenes,[14] calixarenes,[15] branched oligosiloxanes,[16] and H-bonded networks (Chapter 4, this volume; Seto and Whitesides[17]) have each benefitted from the development of synthetic routes that were not too taxing. The design and synthesis of large molecular structures, or *supramolecular synthesis,* has engendered the concepts of *self-assembly* and *nanoscale.* Clearly, what is meant is the creation of molecules with planned features on the scale common in the electronics and other industries, although molecular self-assembly occurs upon crystallization of any molecule or upon addition of surfactants to water above the critical micelle concentration. *Molecular recognition* refers to the design of complexing agents of high specificity, generally with biological antibody–antigen interactions as a model. This area has been reviewed recently by Lehn.[18]

Design of Molecules That Stick Together

Pedersen, in a wonderfully readable review,[19] described how he extended the concept of chelation to two-dimensional rings, the crown ethers, which dramatically increased the solubility of ionic solids in organic solvents.[20,21] Lehn and coworkers have been largely responsible for the extension of complexation of ions to single-molecule structures that chelate in three dimensions.[22] Cram and coworkers have modified the conveniently generated calixarene structure to permit its function as an enzyme,[23] as the host for both willing guests—even the shy xenon atom[24]—and captured guests,[25] and for guests that give birth to cyclobutadiene.[26]

A related and even simpler calixarene derived from resorcinol selectively binds oxygenated molecules as large as simple sugars[27] (Figure 1-6).

The synthesis of large molecules designed for complementary hydrogen bonding has led to a molecular "tape"[28] and to molecules that "recognize" diverse species including 2-aminopyrimidine,[29] *p*-nitrophenol,[30] cyclic adenosine monophosphate,[31] and aliphatic diamines.[32] Recently, Wyler and coworkers reported the synthesis of two molecules that fit together like two parts of the cover of a tennis ball. The stable assembly was able to trap molecules in the interior of the structure.[33] Host molecules have also been designed so that binding of the guest is accompanied by a large increase in fluorescence[34] or alteration in color,[35] as well as differing color responses to one enantiomer of a pair mediated by the pitch of a liquid crystal.[36]

Large molecules have been discovered that combine to form increasingly complex liquid-crystalline supramolecular-polymolecular assemblies,[37] fluid and solid fibers composed of lipid bilayers,[38] lipid tubules,[39] and ion channels,[40,41] and that show nonlinear optical properties,[42] selective electron transfer,[43] and the ability to carry out "signal processing."[44]

The chemical language hes been enriched by the terms used to describe the host in host–guest relationships[45]:

carcerand—host that can encapsulate a guest
podand—an acyclic host, such as terpyridyl
corand—a modified crown ether
perching complex (better, *nesting complex*)
capsular complex—guest surrounded by ligand
spherand—a completely preorganized ligand
spheraplex—a spherand with a guest
cavitand—a concave surface large enough to embrace simple molecules

Noncovalent interactions, among them predominantly hydrogen bonding, are associated with the bulk of biological molecules and their dynamic behavior. The importance of H-bonding in the attachment of

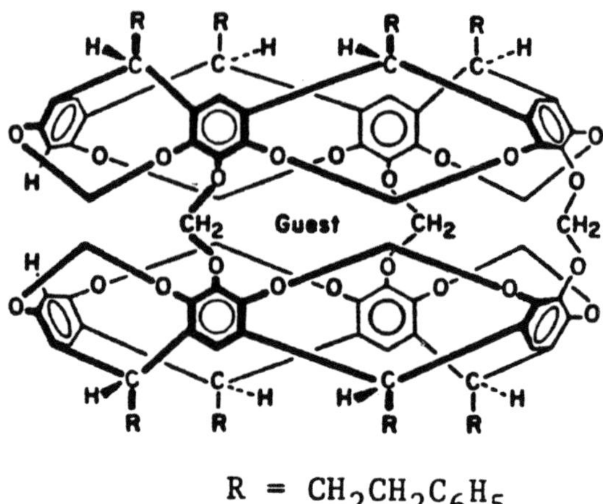

FIGURE 1-6 Calixarene suitable for guest (G) occupancy. [From M.E. Tanner, C.B. Knobler, and D.J. Cram, *J. Am. Chem. Soc.* **112**, 1659 (1990)]

viruses to antibodies and to specific sites on the cell wall has inspired the synthesis of large structures for the inhibition of viral action. Recently, an Australian group, with the aid of computers and a knowledge of the structure of the binding site, has successfully designed molecules to bind with the influenza A virus.[46] Spaltenstein and Whitesides,[47] Weinhold and Knowles,[48] Spevak et al.,[49] and Sabesan[50] and coworkers have constructed clever imitators of the sialicide-containing active site for viral binding. The efficacy of the imitators was enhanced by achieving polyvalency by incorporating them into liposomes or polymerized liposomes[47] (Figure 1-7).

A related concept is the use of large molecules to sterically direct the synthesis of particular stereoisomers, one of which is shown in Figure 1-8. One wonders whether the displacement is S_N2, as portrayed,[51] or S_N2', with attack on the remote double bond position.

The design of three-dimensional H-bonded network polymers is illustrated in detail in Chapter 4 of this text.

Self-replication is one of the more obvious properties of the origins of life. Rebek and coworkers[52–54] synthesized adenine-based nucleotide precursors that, due to selective H-bonding, catalyze their own formation in an interesting echo of the natural process. A similar self-catalysis, in which the favorable thermodynamics of binding of substrate to template enhanced the rate of formation of a Schiff base containing two nucleosides, was described recently by Goodwin and Lynn[55] (Figure 1-9).

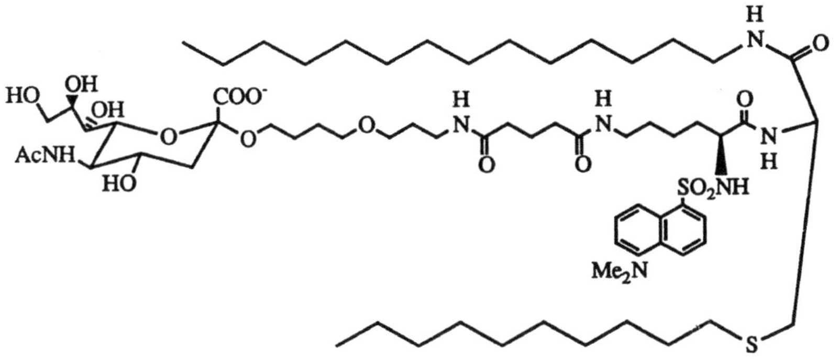

FIGURE 1-7 Fluorescent sialicide that binds to influenza virus. [From E.G. Weinhold and J.R. Knowles, *J. Am. Chem. Soc.* **114,** 9270 (1992)]

Analogous, self-catalyzed condensations of simple molecules are known.[56]

Molecules also may be forced to stick to each other by the trick of knotting them together so that they become topologically linked without covalent bonds. The concept is an old one, but early work in this area, described in very readable detail by Schill,[57] suffered generally from low yields and lengthy syntheses. Dietrich-Buchecker and Sauvage[58] have developed a specialized way to synthesize trefoil and other knots by using metals complexed with chains in the form of a double helix, which yield the topologically intertwined molecules upon removal of the metal. This field has been stimulated by the observation that self-assembled structures are, in some cases, easier to prepare than the independent components.[59]

FIGURE 1-8 Stereospecific alkylation of the derivative of Kemp's acid. [From J.G. Stack, D.P. Curran, S.V. Geib, J. Rebek, Jr., and P. Ballester, *J. Am. Chem. Soc.* **114,** 7007 (1992)]

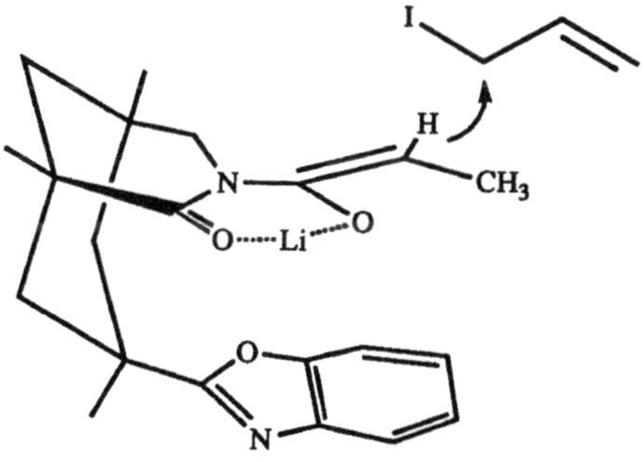

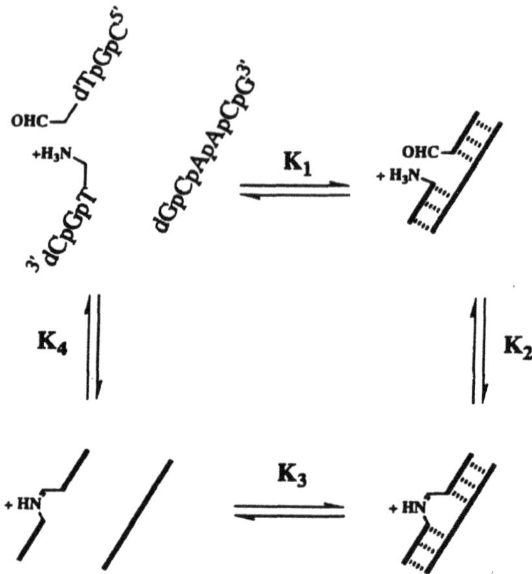

FIGURE 1-9 Catalytic self-replication. [From J.T. Goodwin and D.G. Lynn, *J. Am. Chem. Soc.* **114**, 9197 (1992)]

Splendid examples corresponding to this general scheme in Figure 1-10 are now known.[58]

Self-assembly of linear molecules into cyclodextrins to give threaded loops proceeds in high yields.[60] This approach has been extended to the synthesis of a molecular "train" that moves along a circular aliphatic track from one aromatic ring "station" to another.[61]

Chen and Seeman noted that one of the natural biological forms of DNA was a structure with a right-angle turn. From this, they developed the synthesis of large geometrical structures, including a cube, with approaches perfected in conventional DNA syntheses.[62] Interesting features of this work are the ability to remove by-products enzymatically and the use of Merrifield-type solid supports in the procedure (Figure 1-11).[63]

Molecules that stick together as micelles, bilayer assemblies,[64,65] and liquid crystals[66] represent rather mature but very active disciplines. Recent advances are represented by the syntheses of unusual amphiphiles with mixed fluorocarbon-hydrocarbon chains,[67] extremely long aliphatic chains with multiply ionic head groups,[68] and crown-ether-containing "bola-amphiphiles."[69]

Optical studies of dyes in liquid crystal media as a form of information storage have recently been reviewed.[70]

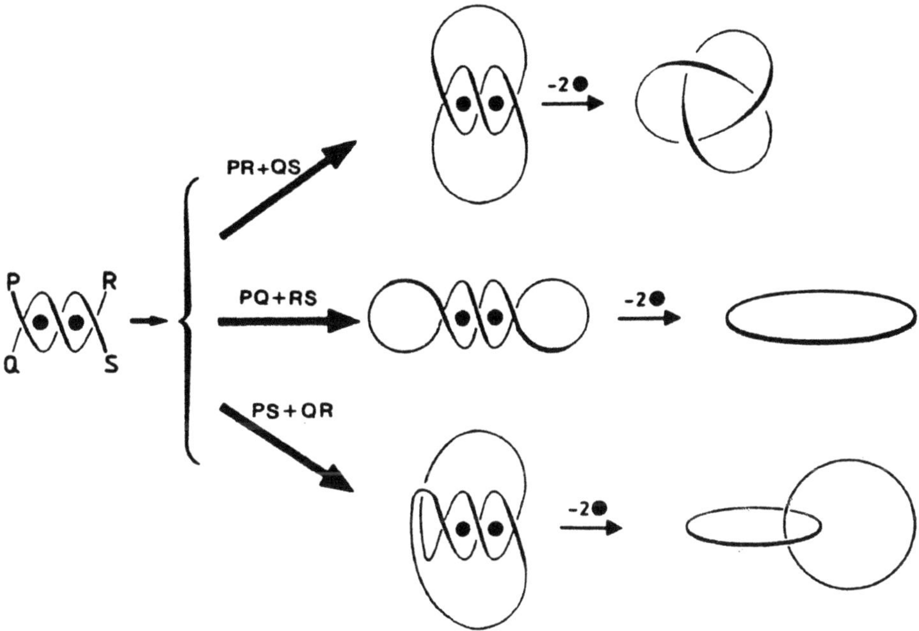

FIGURE 1-10 Derivation of knots from complexes. [From C.O. Dietrich-Buchecker and J.-P. Sauvage, *Ang. Chem. Int. Ed.* **28**, 189 (1989)]

The formation and properties of organized assemblies in the form of gels and foams has come under the scrutiny of fluid dynamicists adept at applying supercomputers to model their behavior.[71] A frequent starting point are space-filling tetrakaidecahedra (truncated octahedra),[72] predicted by Lord Kelvin to be the geometrical shape with the closest resemblance to foam cells.[73]

Aside from traditional, practical applications in processed foods, it is fascinating to discover that hydrocarbon-water gels with large amounts of surfactant are acoustically active and ring in the audible range when struck.[74]

The interplay between natural phenomena and modern technology is represented in studies of that most unusual polymer, spider silk,[75] and efforts to imitate and improve upon its properties with the tools of genetic engineering.[76] Perhaps similar advances will follow from the understanding of muscle action at the molecular level.[77]

Superconductivity

The phenomenon of conductivity traditionally distinguished organic chemistry, with its preponderance of covalently linked structures, from

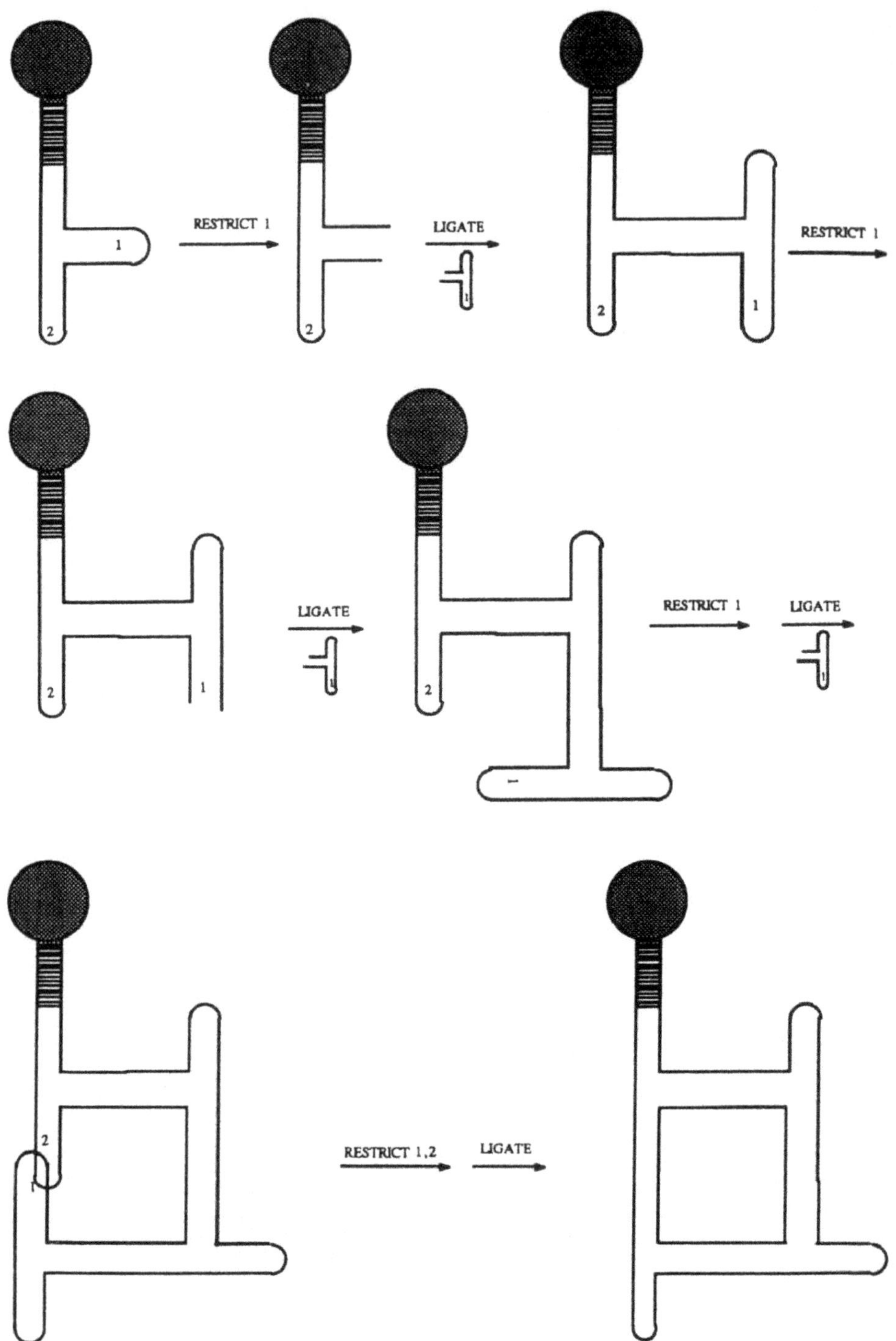

FIGURE 1-11 Geometrical structures from DNA analogues. [From Y. Zhang and N.C. Seeman, *J. Am. Chem. Soc.* **114,** 2656 (1992)]

the high conductivities associated with metals and with solutions of inorganic salts. Indeed, organic polymers made splendid electrical insulation. The development of conductive and, more recently, superconductive organic compounds and polymers has been overshadowed by the spectacular emergence of the superconductive cuprate ceramics,[78] with the current high-temperature record around 130 K for compositions of Hg, Ba, Cu, and O.[79] The organic semi- and superconductor classes of recent interest have been based on small-molecule donor-acceptor complexes with the acceptor a strongly electronegative partner such as tetracyanoethylene (TCNE). Polymeric donors such as poly(phenylene sulfide), poly(aniline), and doped poly(acetylene)[80,81] have been prepared. Poly(acetylene) and numerous derivatives are still under intense scrutiny, reflecting the simplicity of the parent compound, the simplicity of catalytic syntheses, and the potential for increasing its intrinsic stability.[82,83] Similarities in the mechanisms leading to electrical conductivity of doped polymers and those leading to optical nonlinear effects have been pointed out.[84]

The sensitivity of organic conductors to degradation, and probably other factors, have hindered large-scale practical applications in batteries,[85] sensors, and other devices. Organic superconductivities have been achieved only at very low temperatures and face competition from superconducting metals, alloys, and ceramics.

The development of specialized electronic devices and mimicry of miniaturized electronic circuitry with organic-based conductors may be a more promising area.[86]

The study of organic molecular assemblies on metals, particularly gold-thiol systems,[87] may lead to important applications in photoresist and related technology.

The interplay of electronics and chemistry is already richly represented by studies of derivatized electrodes for various purposes,[88,89] analytical methods based on the vibrating quartz microbalance,[90,91] and new combinatorial syntheses of receptors by methods analogous to those of photoresist technology.[92]

Organic Magnets

The chemists' potential contributions to magnetism may be more promising. Traditionally, magnetic fluids[93–95] have been prepared by the brute-force grinding of magnetite or other ferromagnetics for hours with liquid and a surfactant. The process has been improved cleverly by reacting hematite (Fe_2O_3) and CO to give nonmagnetic Wustite (FeO). This softer oxide is ground more easily, and upon heating to 250 °C it disproportionates to give finely divided, ferromagnetic Fe and Fe_3O_4. The magnetic fluids derived from these processes have found many curious specialized applications, such as components of automobile clutches and loudspeakers.[96] A number of magnetite-bound enzymes are available commercially. More recently, the molecular synthesis of ferrofluids from inorganic ions in the presence of natural or synthetic surfactants has been developed.[97]

Hwang[98] has described the combination of a micellar bilayer with a ferromagnetic particle in the center as a potential way to magnetically snare molecules of interest (Figure 1-12). Iron pyrite has been rendered into a magnetically separable form by binding of a specialized ferrofluid to it.[99]

More recently, organic chemists have synthesized single-molecule candidates for ferromagnets, with growing indications of success.[100] A pair of articles by Breslow and colleagues in 1982[101,102] amplified earlier suggestions by McConnell[103] and have stimulated much work in this field.[104]

The most common methods used to prepare organic magnets have centered on polycarbenes,[105] polymerization of acetylene-bearing pendant nitroxides,[106,107] and incorporation of paramagnetic metal ions in crystalline stacks with organic counterions or with coordinating bifunctional ligands of various types.[108] An interesting interpenetrating network assembled from ligands of Mn(II) and Cu(II) became ferromagnetic below 22.5 K.[109] (See also pp. 43–44.)

A bifunctional nitroxide in an adamantanoid framework has been reported to show paramagnetism at 1.4 K.[110]

Sol-gel and Other Syntheses

The molecular approach to ceramic synthesis (Figure 1-13) is well represented in inorganic chemistry. It has interesting commercial applications in the preparation of high-quality, optically transparent components by a colloidal sol-gel process and in lightweight aerogels with superb insulating, optical, and even insecticidal properties.[111,112] The synthesis of organic aerogels derived from melamine- and resorcinol-formaldehyde copolymers has been reviewed.[113] The fractal dimensions of aerogels have been of particular interest to physicists.[114]

More ordered forms of silicon are exemplified by the low-temperature generation of protective silica coatings by oxidation of films of hydrogen silsesquioxane deposited by solvent evaporation.[115]

A recent approach to synthesis at the atomic level is the putative formation of ultrahard carbon nitride from the direct union of C and N atoms in a molecular beam.[116]

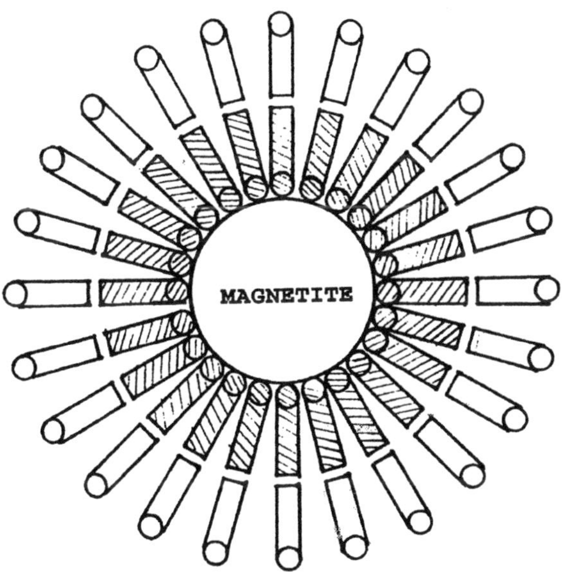

FIGURE 1-12 Idealized structure of surfactant-stabilized magnetic fluid. [From J. Hwang, U.S. Pat. 4,834,898 (1989)]

The Importance of Flaws

The transition from molecules to materials will, with regard to the bulk properties of the solid state, reveal the importance of flaws or defects[117] in determining many important "phenomena such as ionic conduction in crystals, doping effects and p–n junctions in semiconductors, color centers in alkali metal halides, image development in photography, passivation and corrosion of metals, the kinetics of synthesis and sintering of solid materials, problems of rock formation during the earth's evolution, the mechanisms of gas sensors and high-temperature fuel cells, the performance of photosensitive electrodes ... and many more."[118]

Although these concerns have traditionally preoccupied chemists and

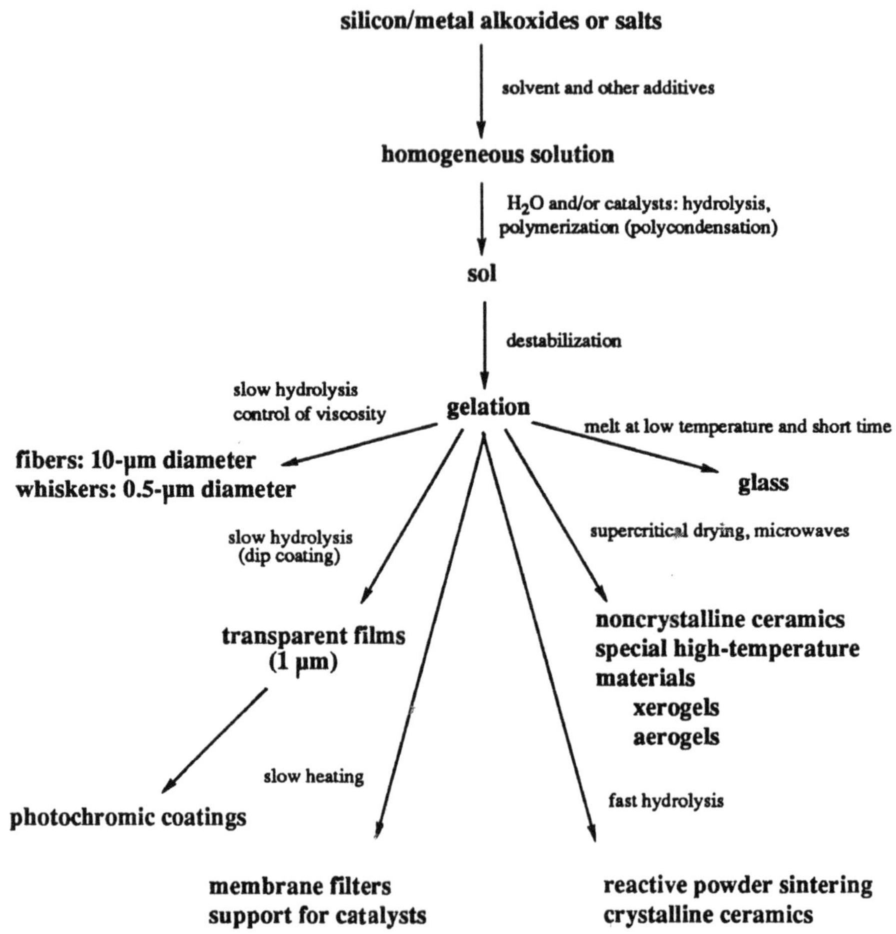

FIGURE 1-13 Synthesis of materials by the sol-gel technique. [From H.D. Gesser and P.C. Goswami, *Chem. Rev.* **89**, 765 (1989)]

physicists in the inorganic and solid-state areas, many of them are pertinent to organic-based materials as well. In addition to the obvious limitations of a material due to the presence of bulk flaws such as cracks (see Chapter 5, this volume) or pinholes in a membrane, there are specific limitations on common organic polymeric materials caused by such defects, for example inaccessible sites in a polymer-bound catalyst. The synthesis of dendrimers using a divergent approach is accompanied by defects from incomplete reaction.[13]

The elimination of defects itself can be the challenge of materials synthesis. For example, crystalline poly(ethylene) approached the theoretical modulus in the aligned direction only after special techniques were developed to perfect the uniformity of chain orientation.[119,120] Once this was accomplished, a new problem arose: it was difficult to find clamps strong enough to hold such aligned samples for testing.

In general, polymeric materials display a range of properties, depending greatly on their mode and temperature of extrusion, cooling history, and degree of crystallinity. The usual representation of semicrystalline structural polymers is one in which long polymeric molecules join together crystalline segments in load-bearing, amorphous zones. The properties of such materials, including simple ones such as density, depend, in turn, upon a number of factors, and will change with time due to chemical reactions and slow physical relaxation processes.

Failure of brittle (ceramic, glassy) materials under stress is treated under crack propagation theory, according to which flaws in the material under cyclic stresses evolve into propagating cracks.[121] The statistical distribution of the flaws in "pure" samples of materials leads to a wide distribution in failure times, making prediction of performance in such materials rather difficult. Even when bulk strength is not a factor, other flaws or even dynamic forms of irregularity (analogous to the problem of magnetic flux lines in ceramic superconductors or protein folding in genetic engineering) may limit the performance of organic materials. While a complete discussion of factors well understood by materials scientists is beyond the scope of this chapter, it may suffice to say that the merger of chemistry with material science involves, for chemists, an entirely new realm of concerns.

A related, classic materials problem is the discontinuity that occurs when dissimilar materials are connected. Although we can hardly do justice to the entire subject of interfacial science in a few sentences, the curious reader can begin by perusing two recent volumes.[122] Researchers in the field of surface chemistry have available a growing number of sophisticated techniques for surface characterization.[123]

To emphasize the diversity of materials research and the fact that significant changes can occur on a larger than molecular scale, consider a study of the "curing" of a low molecular weight, H-bonded runway surface for 40-ton transport planes (Figure 1-14). A composite of this material with sawdust showed enhanced warm-weather stability, but its use was not economically feasible in practice due to the expense of airlifting sawdust.[124]

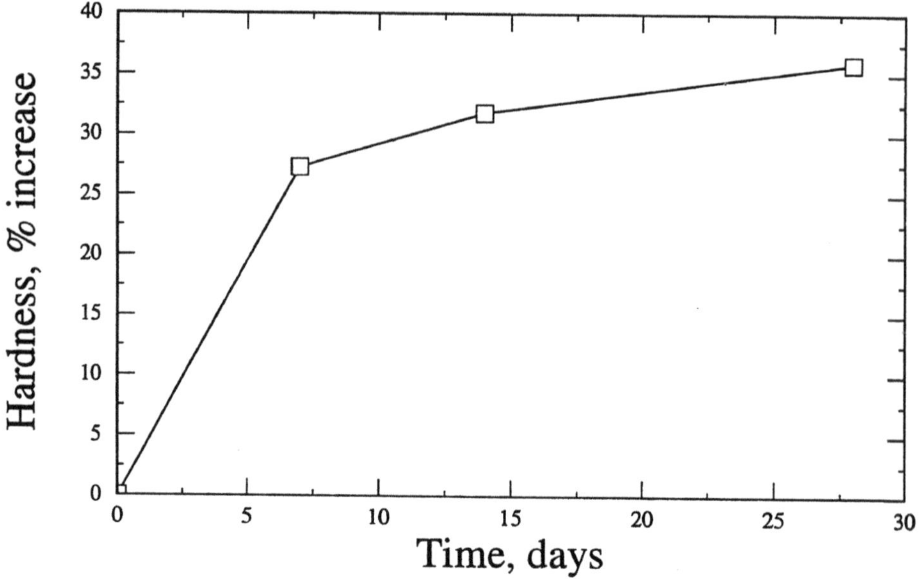

FIGURE 1-14 Experimental study related to landing strips in Antarctica. Force required to indent 3-in.-diameter rod 0.1 in. into packed snow as a function of storage time at −10°C. [Data provided by W. Haas, Department of Civil Engineering, Michigan Technological University]

Thermodynamic Properties of a Large, Volatile Molecule

It is well known that nearly spherical molecules show unusual behavior in the solid state, especially as revealed in phase changes in which rotational motion is possible within a solid lattice (plastic crystals). We searched for such transitions in the new molecule $(t\text{-}BuCH_2CH_2)_4C$, or f(1.2), by calorimetric methods, without success.[125]

Only conventional melting and boiling transitions were observed. Since the heat of the transition from liquid to vapor could not be determined accurately in the calorimetric apparatus, we measured the boiling point of f(1.2) at several temperatures and determined the heat of vaporization from the Clausius-Clapyreon equation. The process was repeated with *n*-hexadecane in order to test our experimental procedure. The experimental data appear in Table 1-1, together with values from a number of other theoretical and experimental approaches.

The differences in the experimental values in the table are much larger than the experimental error for both compounds, and the largest and smallest differ by a factor of nearly 2. Closer examination reveals that the values of ΔH_{vap} in both series decrease at higher temperatures. This is in fact well known, and arises from the different heat capacity of the liquid and gas:

$$-\Delta C_p = \left[\frac{\partial E}{\partial T}\right] + R$$

where

E = energy of vaporization
R = gas constant

The magnitude of this correction and methods to determine it have been reviewed by Shaw.[126] A theoretical calculation of ΔC_p is possible; it requires additional data such as the coefficients of thermal expansion and compressibility. The difference $\Delta C_p = C_p(1) - C_p(g)$ for small molecules was generalized as 10 cal mole^{-1} deg^{-1} (eu) by Tsien.[127]

TABLE 1-1 Calculated and Experimental Values of ΔH_{vap} for f (1.2) and *n*-Hexadecane

Method	T-range, °C	ΔH_{vap} (kcal/mol)
f (1.2)		
Ln (P) vs. $T^{-1(a)}$	262–310	14.0
Trouton's law[b]	310	12.7
Correlation GC[c]	25	[21.7]
Published formula[d]	—	[24.7]
n-$C_{16}H_{34}$		
Ln (P) vs. $T^{-1(a)}$	212–285	14.5
Ln (P) vs. $T^{-1(e)}$	215–285	14.0
Knudsen cell	20–35	19.23 ± 0.08
Evaporation rate	25	19.45 ± 0.08
Trouton's law[b]	285	12.3
Published formula[d]	—	18.6

[a]G.D. Mendenhall, S.X. Liang, and E.H.-T. Chen, *J. Org. Chem.* **55**, 3697 (1990).

[b]F. Trouton, *Phil. Mag.* (London), **18**, 54 (1884). Trouton's law was "discovered" three times between 1823 and 1877: S. Glasstone, *Text-Book of Physical Chemistry* (Van Nostrand, New York, 1940), p. 449.

[c]J.S. Chickos, priv. comm.

[d]J.S. Chickos, R. Annunziata, L.H. Ladon, A.S. Hyman, and J.F. Liebman, *J. Org. Chem.*, **51**, 4313 (1986).

[e]Calculated from data in D L. Camin, A.F. Forziati, and F.D. Rossini, *J. Phys. Chem.*, **58**, 440 (1954).

With the latter value, we predict a correction of only 2–4 kcal/mol between 25 °C and the boiling point of f(1.2) or *n*-hexadecane. Clearly, this is insufficient to account for the variations shown in Table 1-1.

Although these generalizations were expected to deviate somewhat with molecules as large as ours, we quickly resolved the matter when we realized that the known, experimental ΔC_p for the reference compound *n*-hexadecane was an enormous 31.0 eu.[126] The correction for our ΔH_{vap} for *n*-hexadecane is thus calculated to be about 31.0(249 °C − 25 °C) = 6.9 kcal/mol, compared with a difference of 5.0 kcal/mol between experimental values of ΔH_{vap} in Table 1-1 at high vs. low temperatures. This difference of 1.9 kcal/mole is still larger than experimental errors, and it arises because ΔC_p is a value for 25 °C and is itself temperature dependent, declining to zero at the critical temperature.

We assume that the differences in tabulated values for ΔH_{vap} for f(1.2) arise similarly. The experimental number for ΔH_{vap} is interpreted as a (hypothetical) transition between liquid and vapor at 25 °C (21.7 kcal/mol) and is larger than the experimental value near the boiling point (14.0) by an amount corresponding to an average ΔC_p = 29.4 eu, which is very close to the value for *n*-hexadecane.

Given the estimated ΔC_p for f(1.2), we calculate ΔH_{vap} = 16.6 kcal/mol at the melting point, 200 °C. Adding the experimental heat of fusion, 11.6 kcal/mole, we obtain a calculated ΔH_{sub} = 28.2 kcal/mol for temperatures close to the melting point.

The entropy of fusion for f(1.2) is calculated ($\Delta H_{fus}/T_{mp}$) to be 25 eu, which compares favorably to a predicted value of 24.3 eu from empirical relationships in the literature.[128]

The liquid heat capacity of f(1.2) can be estimated as 165.4 eu from additivity relationships,[126] although this figure refers to a hypothetical liquid at 25 °C. The gas-phase heat capacity for various temperatures likewise can be estimated from additivity,[129] and rises from 123 eu at 300 K to 212 eu at 600 K.

Trouton's law appears to be a fairly good predictor for ΔH_{vap} near the boiling point of the compounds in Table 1-1. Other equations relating boiling point to heat of vaporization,[130] applied to f(1.2) and *n*-hexadecane, give results similar to those derived from Trouton's law.

An old equation relating molecular weight and boiling point was derived by McAdams and Morrell[131]:

$$\log_{10} \mathrm{MW} = -4.7523 + 2.5100\ \mathrm{lob}_{10}\ (T_b + 393)$$

The normal boiling point of *n*-hexadecane, 287.5 °C, leads to MW = 228, compared with the actual value of 226. The normal boiling point of f(1.2), 310 °C when inserted into this equation, predicts a molecular weight of 248 compared with the actual value of 353.

The obvious lesson from all of this is that empirical and semiempirical equations are often no more general than the original data base from which they were derived. The properties of very large molecules apparently confound many of these approaches and are therefore valuable to test theoretical models.

Acknowledgment

The author thanks the U.S. Army Research Office (DAAG29-81-K-0077) and Himont U.S.A. for support.

REFERENCES

1. R.B. Seymour and C.E. Carraher, *Polymer Chemistry,* 2nd ed. (Marcel Dekker, Inc., New York, 1988), p. 29.
2. J.M.G. Cowie, *Eur. Polym. J.,* **11,** 297 (1975).
3. G. Schmid, *Chem. Rev.,* **92,** 1709 (1992).
4. M.L. Williams, *J. Phys. Chem.,* **59,** 96 (1955).
5. M.L. Williams, R.F. Landel, and J.D. Ferry, *J. Am. Chem. Soc.,* **77,** 3701 (1955).
6. R.B. Woodward, W.A. Ayer, J.M. Beaton, F. Bickelhaupt, R. Bonnett, P. Buchschacher, G.L. Closs, H. Dutler, J. Hannah, F.P. Hauck, S. Ito, A. Langemann, E. LeGoff, W. Leimgruber, W. Lwowski, J. Sauer, Z. Valenta, and H. Volz, *J. Am. Chem. Soc.,* **82,** 3800 (1960).
7. a. P. Rothemund, *J. Am. Chem. Soc.,* **57,** 2010 (1935). b. J.W. Lehman, *Operational Organic Chemistry,* 2nd ed. (Allyn and Bacon, Boston, 1988), p. 271.
8. J.P. Collman, X. Zhang, R.T. Hembre, and J.I. Brauman, *J. Am. Chem. Soc.,* **112,** 5356 (1990).
9. T.G. Traylor and R. Popovitz-Biro, *J. Am. Chem. Soc.,* **110,** 239 (1988).
10. K. Konishi, K. Oda, K. Nishida, T. Aida, and S. Inoue, *J. Am. Chem. Soc.,* **114,** 1313 (1992).
11. J.T. Groves and R. Neumann, *J. Am. Chem. Soc.,* **109,** 5045 (1987).
12. T.J. Meyer, *Accounts Chem. Res.,* **22,** 163 (1989).
13. D.A. Tomalia, A.M. Naylor, and W.A. Goddard III, *Ang. Chem. Int. Ed.,* **29,** 138 (1990).
14. *Fullerenes* G.S. Hammond, and V.J. Kuck, eds., *ACS Symposium Ser.,* **481** (American Chemical Society, Washington, DC, 1992).
15. a. A. Collet, *Tetrahedron,* **43,** 5725 (1987). b. C.D. Gutsche, *Calixarenes,* in

Monographs in Supramolecular Chemistry, J.F. Stoddart, Series Ed. (Royal Society of Chemistry, Cambridge, 1989).

16. H. Uchida, Y. Kabe, K. Yoshino, A. Kawamata, T. Tsumuraya, and S. Masamune, *J. Am. Chem. Soc.,* **112,** 7077 (1990).
17. C.T. Seto and G.M. Whitesides, *J. Am. Chem. Soc.,* **112,** 6409 (1990).
18. J.M. Lehn, *Ang. Chem. Int. Ed.,* **29,** 1304 (1990).
19. C.J. Pedersen, *Ang. Chem. Int. Ed.,* **27,** 1021 (1988).
20. Aza analogues: a. K.E. Krakowiak, J.S. Bradshaw, and D.J. Zamecka-Krakowiak, *Chem. Rev.,* **89,** 929 (1989). b. J.S. Bradshaw, K.E. Krakowiak, and R.M. Izatt, *Aza-crown Macrocycles* (John Wiley & Sons, New York, 1993).
21. Sulfur analogues: S.R. Cooper, *Acc. Chem. Res.,* **21,** 141 (1988).
22. J.M. Lehn, *Ang. Chem. Int. Ed.,* **27,** 1095 (1988). Recent review: C. Seel and F. Vogtle, *Ang. Chem. Int. Ed.,* **31,** 528 (1992).
23. D.J. Cram, *Ang. Chem. Int. Ed.,* **27,** 1009 (1988).
24. M.E. Tanner, C.B. Knobler, and D.J. Cram, *J. Am. Chem. Soc.,* **112,** 1659 (1990).
25. J.A. Bryant, M.T. Blanda, M. Vincenti, and D.J. Cram, *J. Am. Chem. Soc.,* **113,** 2167 (1991).
26. D.J. Cram, M.E. Tanner, and R. Thomas, *Ang. Chem. Int. Ed.,* **30,** 1024 (1991).
27. a. Y. Tanaka, C. Khare, M. Yonezawa, and Y. Aoyama, *Tetrahedron Lett.,* **31,** 6139 (1990). b. Y. Aoyama, Y. Tanaka, and S. Sugahara, *J. Am. Chem. Soc.,* **111,** 5397 (1989).
28. J.A. Zerkowski, C.T. Seto, D.A. Wierda, and G.M. Whitesides, *J. Am. Chem. Soc.,* **112,** 9025 (1990).
29. J.C. Adrian, Jr., and C.S Wilcox, *J. Am. Chem. Soc.,* **113,** 678 (1991).
30. J.E. Cochran, T.J. Parrott, B.J. Whitlock, and H.W. Whitlock, *J. Am. Chem. Soc.,* **114,** 2269 (1992).
31. G. Deslongchamps, A. Galan, J. de Mendoza, and J. Rebek, Jr., *Ang. Chem. Int. Ed. Engl.,* **31,** 61 (1992).
32. F. Ebmeyer and J. Rebek, Jr., *Ang. Chem. Int. Ed.,* **29,** 1149 (1990).
33. R. Wyler, J. de Mendoza, and J. Rebek, Jr., *Ang. Chem. Int. Ed.,* **32,** 1699 (1993).
34. A. Prasanna de Silva and K.R.A. Samankumara Sandanayake, *Ang. Chem. Int. Ed.,* **29,** 1173 (1990).
35. H.-G. Lohr and F. Vogtle, *Ang. Chem. Int. Ed.,* **30,** 958 (1991).
36. T. Nishi, A. Ikeda, T. Matsuda, and S. Shinkai, *J. Chem. Soc. Chem. Commun.,* 339 (1991).
37. T. Gulik-Krzywicki, C. Fouquey, and J.M. Lehn, *Proc. Natl. Acad. Sci. U.S.A.,* **90,** 163 (1993).
38. J.-H. Fuhrhop and W. Helfrich, *Chem. Rev.,* **93,** 1565 (1993).

39. J.M. Schnur, *Science,* **262,** 1669 (1993).
40. A. Nakano, Q. Xie, J.V. Mallen, L. Echegoyen, and G.W. Gokel, *J. Am. Chem. Soc.,* **112,** 1287 (1990).
41. J. Canceill, L. Jullien, L. Lacombe, and J.M. Lehn, *Helv. Chim. Acta,* **75,** 791 (1992).
42. J.-M. Lehn, in *Materials for Nonlinear Optics,* S.R. Marder, J.E. Sohn, and G.D. Stucky, eds., *ACS Symp. Ser.,* **455** (American Chemical Society, Washington, DC, 1991), ch. 28.
43. N.J. Turro, J.K. Barton, and D.A. Tomalia, *Acc. Chem. Res.,* **24,** 332 (1991).
44. J.M. Lehn, in *Front. Supramol. Org. Chem. Photochem.,* H.J. Schneider and H. Duerr, Eds. (VCH, Weinheim, 1991), pp. 1–28.
45. H.-J. Schneider, *Ang. Chem. Int. Ed.,* **30,** 1417 (1991).
46. M. von Itzstein, W.-Y. Wu, G.B. Kok, M.S. Pegg, J.C. Dyason, B. Jin, T.V. Phan, M.L. Smythe, H.F. White, S.W. Oliver, P.M. Colman, J.N. Varghese, D.M. Ryan, J.M. Cameron, and C.R. Penn, *Nature,* **363,** 418 (1993).
47. A. Spaltenstein and G.M. Whitesides, *J. Am. Chem. Soc.,* **113,** 686 (1991).
48. E.G. Weinhold and J.R. Knowles, *J. Am. Chem. Soc.,* **114,** 9270 (1992).
49. W. Spevak, J.O. Nagy, D.H. Charych, M.E. Schaefer, J.H. Gilbert, and M.D. Bednarski, *J. Am. Chem. Soc.,* **115,** 1146 (1993).
50. S. Sabesan, J.O. Duus, S. Neira, P. Domaille, S. Kelm, J.C. Paulson, and K. Bock, *J. Am. Chem. Soc.,* **114,** 8363 (1992).
51. J.G. Stack, D.P. Curran, S.V. Geib, J. Rebek, Jr., and P. Ballester, *J. Am. Chem. Soc.,* **114,** 7007 (1992).
52. J. Rebek, Jr., *Acc. Chem. Res.,* **23,** 399 (1990).
53. T. Tjivikua, P. Ballester, and J. Rebek, Jr., *J. Am. Chem. Soc.,* **112,** 1249 (1990).
54. J.-I. Hong, Q. Feng, V. Rotello, and J. Rebek, Jr., *Science,* **255,** 848 (1992).
55. J.T. Goodwin and D.G. Lynn, *J. Am. Chem. Soc.,* **114,** 9197 (1992).
56. A. Terfort and G. von Kiedrowski, *Ang. Chem. Int. Ed.,* **31,** 654 (1992).
57. G. Schill, *Catenanes, Rotaxanes, and Knots* (Academic Press, New York, 1971).
58. C.O. Dietrich-Buchecker and J.-P. Sauvage, *Ang. Chem. Int. Ed.,* **28,** 189 (1989); *Chem. Rev.,* **87,** 795 (1987).
59. P.R. Ashton, A.S. Reder, N. Spencer, and J.F. Stoddart, *J. Am. Chem. Soc.,* **115,** 5286 (1993).
60. T.V.S. Rao and D.S. Lawrence, *J. Am. Chem. Soc.,* **112,** 3614 (1990).
61. P.R. Ashton, C.L. Brown, E.J.T. Chrystal, K.P. Parry, M. Pietraszkiewicz, N. Spencer, and J. Fraser-Stoddart, *Ang. Chem. Int. Ed.,* **30,** 1043 (1991).
62. J. Chen and N.C. Seeman, *Nature,* **350,** 631 (1991).
63. J.-H. Chen, N.R. Kallenbach, and N.C. Seeman, *J. Am. Chem. Soc.,* **111,** 6402 (1989).

64. T. Kunitake, *Ang. Chem. Int. Ed.*, **31**, 709 (1992).
65. F.M. Menger, *Ang. Chem. Int. Ed.*, **30**, 1086 (1991).
66. *Recent Advances in Liquid Crystalline Polymers*, L.L. Chapoy, Ed., *Proceedings of the 6th European Polymer Science Workshop* (Elsevier, New York, 1985).
67. H. Kuwahara, M. Hamada, Y. Ishikawa, and T. Kunitake, *J. Am. Chem. Soc.*, **115**, 3002 (1993).
68. F.M. Menger and Y. Yamasaki, *J. Am. Chem. Soc.*, **115**, 3840 (1993).
69. S. Munoz, J. Mallen, A. Nakano, Z. Chen, I. Gay, L. Echegoyen, and G.W. Gokel, *J. Am. Chem. Soc.*, **115**, 1705 (1993).
70. H.-W. Schmidt, *Ang. Chem. Int. Ed.*, **28**, 940 (1989).
71. a. D.A. Reinelt and A.M. Kraynik, *J. Colloid Interface Sci.*, **159**, 460 (1993). b. D.A. Reinelt and A.M. Kraynik, *J. Fluid Mech.*, **215**, 431 (1990).
72. R. Williams, *The Geometrical Foundation of Natural Structure* (Dover Publications, New York, 1979), p. 167.
73. W. Thompson (Lord Kelvin), *Phil. Mag.*, **24**, 503 (1887).
74. H. Hoffmann and G. Ebert, *Ang. Chem. Int. Ed.*, **27**, 902 (1988).
75. R.V. Lewis, *Acc. Chem. Res.*, **25**, 392 (1992).
76. a. Silk polymers: *Materials Science and Biotechnology*, D. Kaplan, W.W. Adams, B. Farmer, and C. Viney, Eds., *ACS Symp. Ser.*, **544** (American Chemical Society, Washington, DC, 1994). b. Review: J. Cappello, *MRS Bull.*, **17**, 48 (1992).
77. F.W. Taylor, *Science*, **261**, 35 (1993) and references therein.
78. C.P. Poole, Jr., T. Datta, and H.A. Farach, *Copper Oxide Superconductors* (Wiley, New York, 1988)
79. A. Schilling, M. Cantoni, J.D. Guo, and H.R. Ott, *Nature*, **363**, 56 (1993).
80. J.M. Williams, M.A. Beno, H.H. Weng, P.C.W. Leung, T.J. Emge, U. Geiser, and K.D. Carlson, *Acc. Chem. Res.*, **18** 261 (1985).
81. H. Inokuchi, *Ang. Chem. Int. Ed.*, **27**, 1747 (1988).
82. B. Chu and R. Xu, *Acc. Chem. Res.*, **24**, 384 (1991).
83. N.S. Sariciftci, V.M. Kobryanskii, M. Reghu, L. Smilowitz, C. Halvorson, T.W. Hagler, D. Mihailovic, and A.J. Heeger, *Synth. Met.*, **53**, 161 (1993).
84. J.L. Bredas, F. Meyers, and A.J. Heeger, *NATO ASI Ser. E*, **194**, 23 (1991).
85. T. Shimidzu, *AIP Conf. Proc.*, **262**, 129 (1992).
86. R.H. Baughman, *Makromol. Chem., Macromol. Symp.*, **51**, 193 (1991).
87. L.H. Dubois and R.G. Nuzzo, *Ann. Rev. Phys. Chem.*, **43**, 437 (1992).
88. *Electrochemical Surface Science: Molecular Phenomena at Electrode Surfaces*, M.P. Soriaga, Ed., *ACS Symp. Serv.*, **378** (American Chemical Society, Washington, DC, 1988).
89. *Chemically Modified Surfaces in Catalysis and Electrocatalysis*, J.S. Miller,

Ed., *ACS Symp. Ser.*, **192** (American Chemical Society, Washington, DC, 1982).

90. a. D.A. Buttry and M.D. Ward, *Chem. Rev.*, **92,** 1355 (1992). b. M.D. Ward and D.A. Buttry, *Science*, **249,** 1000 (1990).
91. *Applied Piezoelectric Quartz Crystal Microbalances*, Vol. **7,** C. Lu and A. Czanderna, Eds. (Elsevier Sequoia, Lausanne, 1984).
92. A. Borchardt and W.C. Still, *J. Am. Chem. Soc.*, **116,** 373 (1994).
93. R.E. Rosensweig, *Sci. Am.*, **247,** 136 (1982).
94. T.A. Witten, *Physics Today*, July 1990, p. 21.
95. S.E. Khalafalla, *Chemtech*, September 1975, p. 540.
96. K. Raj and R.J. Moskowitz, *Magnetism and Mag. Materials*, **85,** 233 (1990).
97. S.W. Charles and J. Popplewell, *IEEE Trans. Magnetics*, MAG-16, No. 2, March 1980.
98. J. Hwang, U.S. Pat. 4,834,898 (1989).
99. J.Y. Hwang, *Magnetic Reagent Technology for Mineral Processing*, Society of Mining Engineers Annual Meeting, February 1992, Pheonix, AZ, SME Preprint 92-226.
100. P. Day, *Science*, **261,** 402 (1993).
101. R. Breslow, B. Jaun, R.Q. Kluttz, and C-Z. Xia, *Tetrahedron*, **38,** 863 (1982).
102. R. Breslow, *Pure Appl. Chem.*, **54,** 927 (1982).
103. H.M. McConnell, *J. Chem. Phys.*, **39,** 1910 (1963).
104. C. Kollmar and O. Kahn, *Acc. Chem. Res.*, **26,** 259 (1993).
105. H. Iwamura and N. Koga, *Acc. Chem. Res.*, **26,** 346 (1993).
106. J.S. Miller, A.J. Epstein, and W.M. Reiff, *Acc. Chem. Res.*, **21,** 114 (1988).
107. K. Inoue, N. Koga, and H. Iwamura, *J. Am. Chem. Soc.*, **113,** 9803 (1991).
108. A. Caneschi, D. Gatteschi, R. Sessoli, and P. Rey, *Acc. Chem. Res.*, **22,** 392 (1989).
109. H.O. Stumpf, L. Ouahab, Y. Pei, D. Grandjean, and O. Kahn, *Science*, **261,** 447 (1993).
110. R. Chiarelli, M.A. Novak, A. Rassat, and J.L. Tholence, *Nature*, **363,** 147 (1993).
111. R.D. Shoup, in *Ultrastructure Processing of Advanced Ceramics*, J.D. Mackenzie and D.R. Ulrich, Eds. (Wiley, New York, 1988), p. 347.
112. *Chemical Processing of Advanced Materials*, L.L. Hench and J.K. West, Eds. (Wiley-Interscience, New York, 1992).
113. H.D. Gesser and P.C. Goswami, *Chem. Rev.*, **89,** 765 (1989).
114. E. Courtens and R. Vacher, *Philos. Mag.*, **B.65,** 347 (1992).
115. G. Chandra, *Mat. Res. Soc. Symp. Proc.*, **203,** 97 (1991).
116. C. Niu, Y.Z. Lu, and C.M. Lieber, *Science*, **261,** 334 (1993).
117. *Defects in Materials*, Symposium of the Materials Research Society, P.D. Bristowe et al., Eds. (MRS, Pittsburgh, 1991).

118. J. Maier, *Ang. Chem. Int. Ed.,* **32,** 313 (1993); see also p. 528.

119. T. Kanamoto, A. Tsuruta, K. Tanaka, M. Takeda, and R.S. Porter, *Polym. J.,* **15,** 327 (1983).

120. L.-H Wang and R.S. Porter, *J. Poly. Sci. B: Polym. Phys.,* **28,** 2441 (1990).

121. T.L. Anderson, *Fracture Mechanics: Fundamentals and Applications* (CRC Press, Boca Raton, FL, 1991), ch. 8.

122. a. *Interfaces Between Polymers, Metals, and Ceramics,* B.M. DeKoven, A.J. Gellman, and R. Rosenberg, Eds., *Materials Research Society Symposia Proceedings,* **153** (MRS, Pittsburgh, 1989). b. *Materials Interfaces,* D. Wolf and S. Yip, Eds. (Chapman and Hall, New York, 1992).

123. T.J. Shaffner, *Surf. Interface Anal.,* **14,** 598 (1989).

124. W. Haas, priv. comm.

125. G.D. Mendenhall, S.X. Liang, and E.H.-T. Chen, *J. Org. Chem.,* **55,** 3697 (1990).

126. R. Shaw, *J. Chem. Eng. Data,* **14,** 461 (1969). See also J.S. Chickos, D.G. Hesse, S.Y. Panshin, D.W. Rogers, M. Saunders, P.M. Uffer, and J.F. Liebman, *J. Org. Chem.,* **57,** 1897 (1992).

127. H.S. Tsien, *J. Am. Rocket Soc.,* **23,** 17 (1953).

128. J.S. Chickos, D.G. Hesse, and J.F. Liebman, *J. Org. Chem.,* **55,** 3833 (1990).

129. S.W. Benson, *Thermochemical Kinetics,* 2nd ed. (Wiley-Interscience, New York, 1976), p. 272.

130. W. Kistiakowsky, *Z. Physik. Chem.,* **B107,** 65 (1923).

131. W.H. McAdams and J.C. Morrell, *Ind. Eng. Chem.,* **16,** 375 (1924).

2

Cascade Molecules

GEORGE R. NEWKOME AND
CHARLES N. MOOREFIELD

INTRODUCTION

In 1986 Menger[1] published an excellent review of multiarmed organic molecules that included polyethers, poly-armed molecules possessing either ionic or neutrally terminated chains, and some conventionally prepared branched polymers. Over the years, interesting names have been applied to many of these molecules, such as *tripod, supertripod, tetrapod, octopus, hexahost, polypod, hexapus* and *tentacle,* in an attempt to describe these complex structures.

During the period since Menger's review appeared, the number of reported syntheses and utilization of multiarmed organic molecules has exploded and will undoubtedly continue to do so, since these polyfunctional molecules afford ideal entrance[2–7] to supramolecular chemistry.[8,9] In this chapter, we focus on their utilization as core building blocks in order to afford structural directionality and hence define initial parameters, such as internal and external functional group density, for cascade polymers or dendritic macromolecules. To envision this concept, the *cascade molecule* (e.g., multiarm organic compounds) will be the initial building block for subsequent polymerization in a specific tiered or stepwise sequence.

Due to space limitations, this chapter does not cover all of the fascinating cascade molecules that have recently been reported. For the most part, it is limited to those appearing since Menger's presentation.

ARBOROLS, CASCADES, AND DENDRIMERS: THE EARLY YEARS

In our laboratories,[10] we devised a new route to multiarmed molecules, which were subsequently transformed into macromolecules (cascade polymers) via sequential tier development (Scheme 2-1). It is noteworthy that in these early attempts, the inherent neopentyl substitution pattern caused by the limited choice of building blocks was difficult to circumvent under reasonable reaction conditions;[11] thus, homologation procedures were employed[12] to address this problem. In these divergent synthetic procedures, all steps had to be straightforward, since as the number of reaction sites increased with each tier, facile conversion under reasonable reaction conditions was essential and all steps had to be in high yields. Thus, when an alkyl halide was treated with trialkyl methanetricarboxylate,[13,14] the desired triester was obtained; however, attempted reduction to the corresponding triol was plagued with fragmentation products. The desired first-tier triol (**1**) was obtained by reaction of the appropriate aldehyde with excess formaldehyde in the presence of base. A spacer moiety was subsequently incorporated to circumvent the inherent neopentyl deactivation to nucleophilic substitution to give triester **2**. To complete the sequence, reduction afforded the homologated triol that was tosylated and smoothly converted to the nonaester (**3** and **4**, respectively), followed by transformation with Tris (trihydroxymethylaminomethane) to [27]-arborol **5**.

The name [27]-arborol is derived from [27], denoting the number of surface functional groups; *arbor,* derived from the Latin word for "tree," and *ol,* derived from the surface functionality, in this case an alcohol. Due to the complexity of these and related cascade molecules, numerous such logos have been devised; a new, systematic "cascade" nomenclature has recently been proposed,[15] affording a logical, uniform communication pathway to researchers interested in this field (see also p. 181).

Utilizing the above simple triester-Tris conversion sequence, the two-directional arborols were readily formed. The initial discovery[16] that arborols from the [9]-$(CH_2)_n$-[9] and [6]-$(CH_2)_n$-[6] series ([ ] denotes the number of surface hydroxy groups; Figure 2-1) formed aqueous gels via dumbbell-like stacking provided the impetus for further study[17] of these unique molecules. Since the lipophilic cores were layered orthogonally on top of each other, it was surmised that incorporation of unsaturated sites, centrally positioned in the core, would potentially allow the gelled aggregates to be polymerized via this preorganization stacking process. Therefore, the two-directional arborols containing either a centrally located, bridging triple or double bond were constructed (Scheme 2-2).

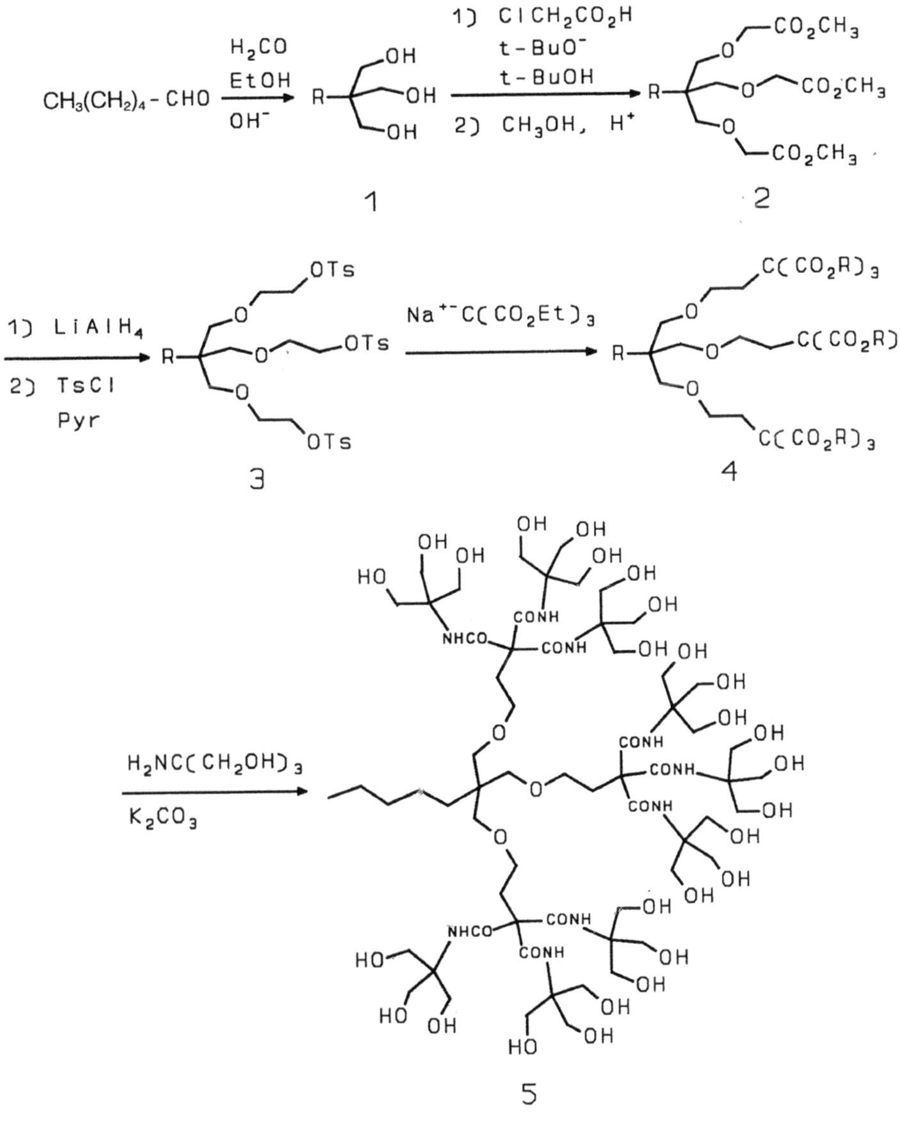

$CH_3(CH_2)_4$-CHO $\xrightarrow[OH^-]{H_2CO,\ EtOH}$ R–C(CH_2OH)$_3$ **1** $\xrightarrow[2)\ CH_3OH,\ H^+]{1)\ ClCH_2CO_2H,\ t\text{-}BuO^-,\ t\text{-}BuOH}$ R–C($CH_2OCH_2CO_2CH_3$)$_3$ **2**

$\xrightarrow[2)\ TsCl,\ Pyr]{1)\ LiAlH_4}$ R–C($CH_2OCH_2CH_2$OTs)$_3$ **3** $\xrightarrow{Na^{+-}C(CO_2Et)_3}$ R–C($CH_2OCH_2CH_2C(CO_2R)_3$)$_3$ **4**

$\xrightarrow[K_2CO_3]{H_2NC(CH_2OH)_3}$ **5**

SCHEME 2-1 Preparation of a [27]-arborol using a spacer moiety to allow facile construction of intermediate nonaester **4**.

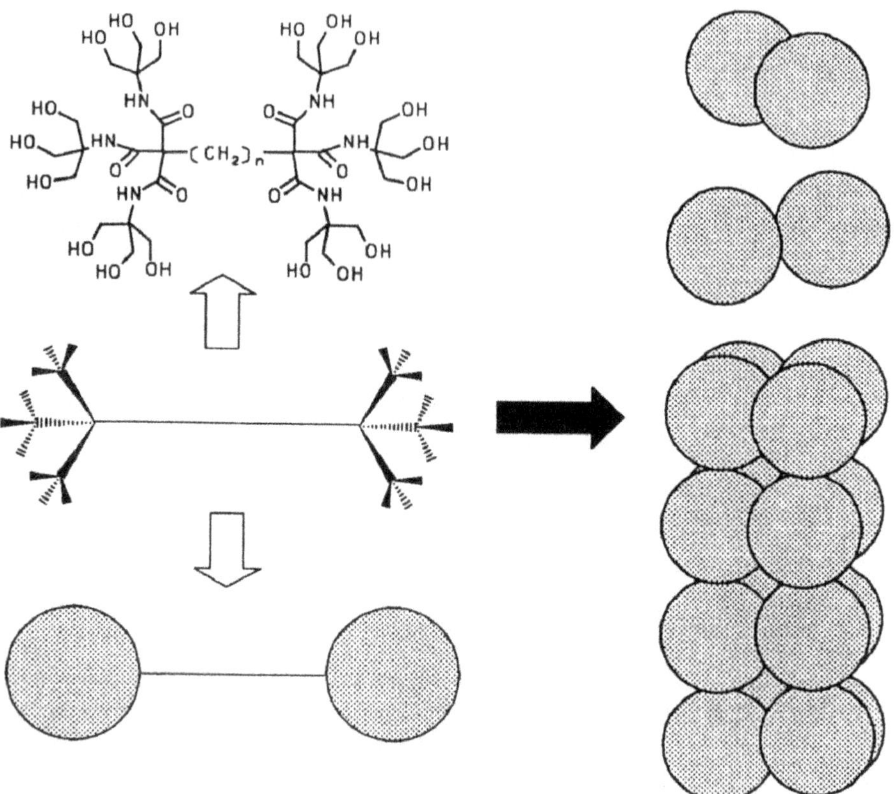

FIGURE 2-1 Proposed dumbbell-like stacking of [9]-$(CH_2)_n$-[9] arborols, where $n = 10$ or greater.

Preparation of the *alkyne* arborols (**6-8**) was readily accomplished[18] by starting with an appropriate chloroalkanol, which was first protected with dihydropyran yielding a chlorotetrahydropyranyl ether and then transformed into a terminal alkyne via treatment with lithium acetylide. Deprotection of the alcohol moieties, mesylation, reaction with "triester" $[HC(CO_2Et)_3]$, and treatment with Tris afforded the one-directional, terminal alkyne arborol **6.** A Cu-promoted coupling of two equivalents of **6** to give dialkyne arborol **7** was unsuccessful, but coupling of the initial terminal alkyne tetrahydropyranyl (THP) ether proceeded smoothly to give the two-directional *di*alkyne core, which was then transformed, in several steps (see Scheme 2-2) to yield the $[9](CH_2)_6(C{\equiv}C)_2(CH_2)_6[9]$ *di*alkyne arborol **7.**

A desired two-directional monoalkyne core was constructed via C-C bond formation with a lithium acetylide THP ether and the correspond-

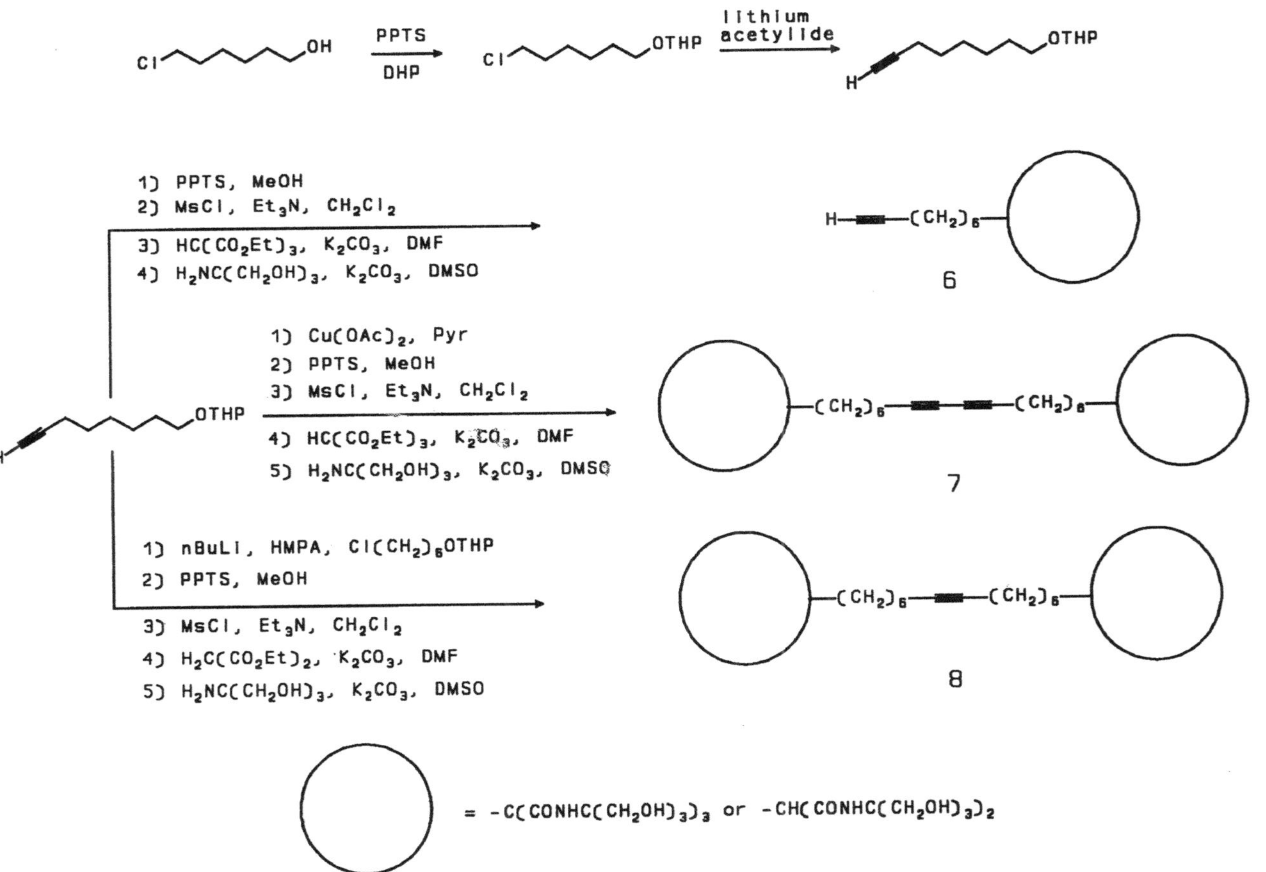

SCHEME 2-2 Syntheses of alkyne arborols.

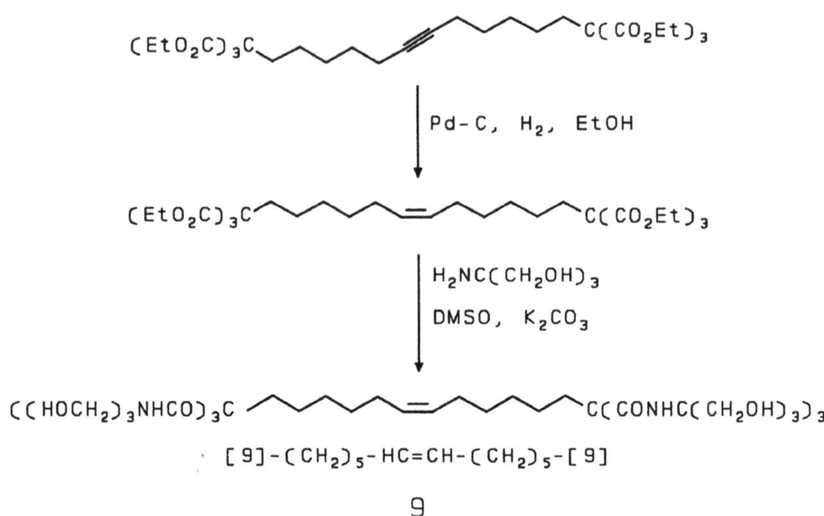

SCHEME 2-3 Alkene arborols prepared by reduction of alkyne arborol intermediates.

ing chloro THP ether. The resultant bis(THP) ether was subsequently deprotected, converted to a dimesylate, reacted with diethyl malonate, and finally treated with Tris in DMSO, affording [6]-$(CH_2)_6C{\equiv}C(CH_2)_6$-[6] arborol **6.**

The [9]-$(CH_2)_5CH{=}CH(CH_2)_5$-[9] arborol **9** was obtained by a similar procedure (Scheme 2-3), except that before final amidation with Tris, the triple bond was partially reduced to a double bond via the use of a Lindlar catalyst.

Catalyzed polymerization of the aqueous gels derived from these alkene and alkyne arborols has thus far been unsuccessful; however, electron microscopy of the [6]-$(CH_2)_6C{\equiv}C(CH_2)_6$-[6] gel not only confirmed the stacked aggregation but also revealed another interesting phenomenon. This alkyne aggregate possessed a unique helical morphology (Figure 2-2), presumably promoted by the increased core rigidity and electron repulsion of the juxtaposed alkynes, preventing the true orthogonal configuration denoted in the related alkane series. The large diameters of the twisted rods (≈600 Å) demonstrated in the electron micrographs probably result from the packing of individual rods into the grooves of adjacent helical rods, or aggregates, thereby producing "supracoil" or "molecular rope" nanostructures.[18] These cascade molecules therefore possess the ability to "self-assemble" into a helix much like that of actin filaments that bundle into supramolecular structures such as stress fibers,[19] acrosomal processes,[20] microvilli,[21] and adhesion belts.[22]

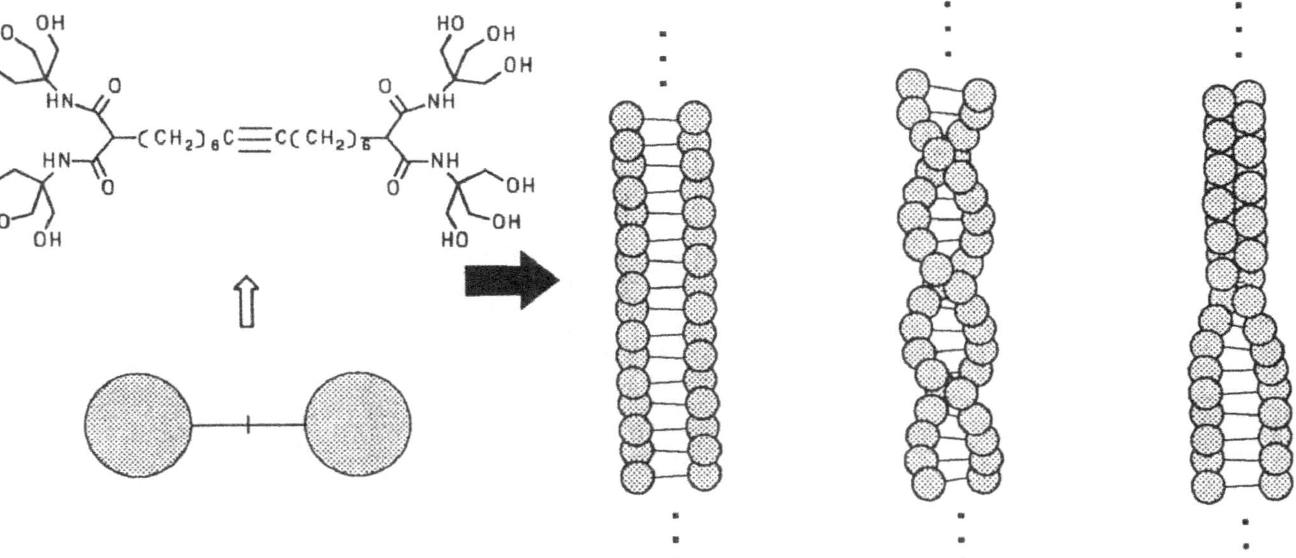

FIGURE 2-2 Postulated aggregation of the alkyne arborols. Association of the individual aggregates, or stacks, leads to "molecular ropes."

CASCADOLS: POLYETHEREAL ALCOHOLS

The synthesis of a four-directional "Cascadol" has recently been reported.[23] The preparative strategy involved the use of a methane C-core, prepared by homologation of pentaerythritol to give a polyether tetraol (Scheme 2-4). Treatment of the tetraol core with dimsyl anion in DMSO, followed by addition of a novel tris(trityl)monomesylate tetraol building block, afforded (16%) the dodecatrityl ether. Removal of the trityl protecting groups and subsequent acetylation yielded cascade dodecaacetate **10.**

POLYAMINES

McAuley et al.[24-26] prepared dendritic polyamines to be used for the examination of Ni complexes. Small, highly compact tetraamine **11** was synthesized from pentaerythrityl tetrabromide[27] upon treatment with a salt of *p*-toluenesulfonamide, under drastic conditions in view of the above-noted nucleophilic inertness instilled by the neopentyl moiety, followed by deprotection of the tetratosylate (Scheme 2-5). Polyamines **12** and **13** were similarly prepared,[26,28] using 1,2-ethylenediamine, and were isolated from the same reaction mixture in 50-60% and 2-5%, respectively. Isolation and purification of **12** and **13** involved the initial formation of Ni complexes.

ETHEREAL POLYPODANDS

Related polyethereal chains have been derived[29] from the readily available 2-alkyl-1,3-propanediols and 2,2-*bis*(hydroxymethyl)alkanols (Scheme 2-6) via initial homologation similar to that of the above arborol examples. The di- or triols, when treated with base and chloroacetic acid, followed by reduction with lithium aluminum hydride, afforded etherol side chains (not depicted), whereas tosylation, followed by addition of the sodium salt of an ethereal capped glycol, afforded the dipodands **14** and tripodands **15,** respectively. The hexadentate thioether ligands were also prepared[30,31] from the intermediary tritosylate, despite the inherent generally low reactivity of these tosylates. Treatment with three equivalents of sodium 2-methylthioethanethiolate under reflux in a steel tube afforded reasonable yields of the desired ligand **16;** the related hexaamines were prepared in a similar way.[32]

The introduction of chirality into *tris*(hydroxymethyl)methanes[33] offers an interesting entrance to a related series of chiral cascade molecules.

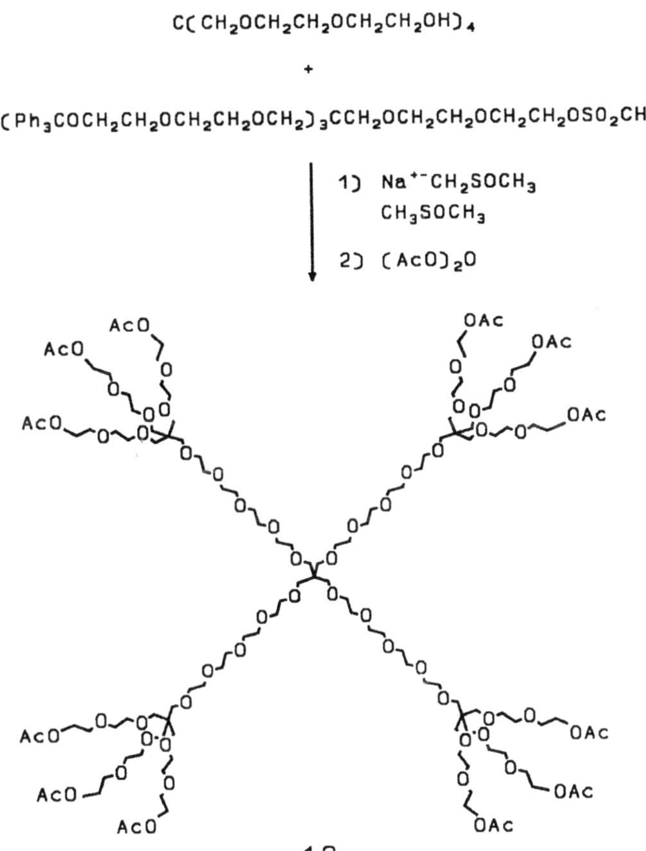

SCHEME 2-4 Synthesis of the polyether (**10**), "Cascadol."

These simple starting materials are ideal core candidates to attach amino acids and peptides, as recently demonstrated.[34] Interestingly, coupling conditions using DCC or DCC-hydroxybenzotriazole failed to afford the desired peptide bond, but another activated ester method (formation of the *p*-nitrophenyl ester **17** using DCC/4-dimethylaminopyridine) worked in high yields. These activated ester intermediates were treated with the methyl esters of glycine (and glycineglycine, not shown) to afford the desired amino acid derivatives, **18.**

An improved synthesis[35] of the simplest triol starting material, 2-hydroxymethyl-1,3-propanediol, and its derivatives from 2-hydroxymethyl-2-nitro-1,3-propanediol has been realized by initial benzylation, replacement of the nitro group with hydrogen by treatment with tri-*n*-butylstannane, and

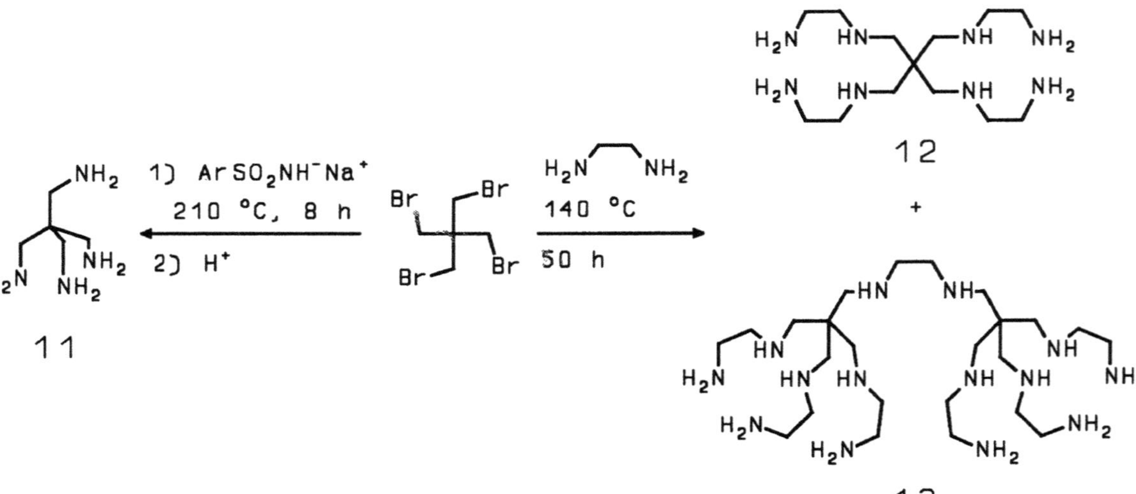

SCHEME 2-5 Polyamine construction employing pentaerythrityl tetrabromide as a core.

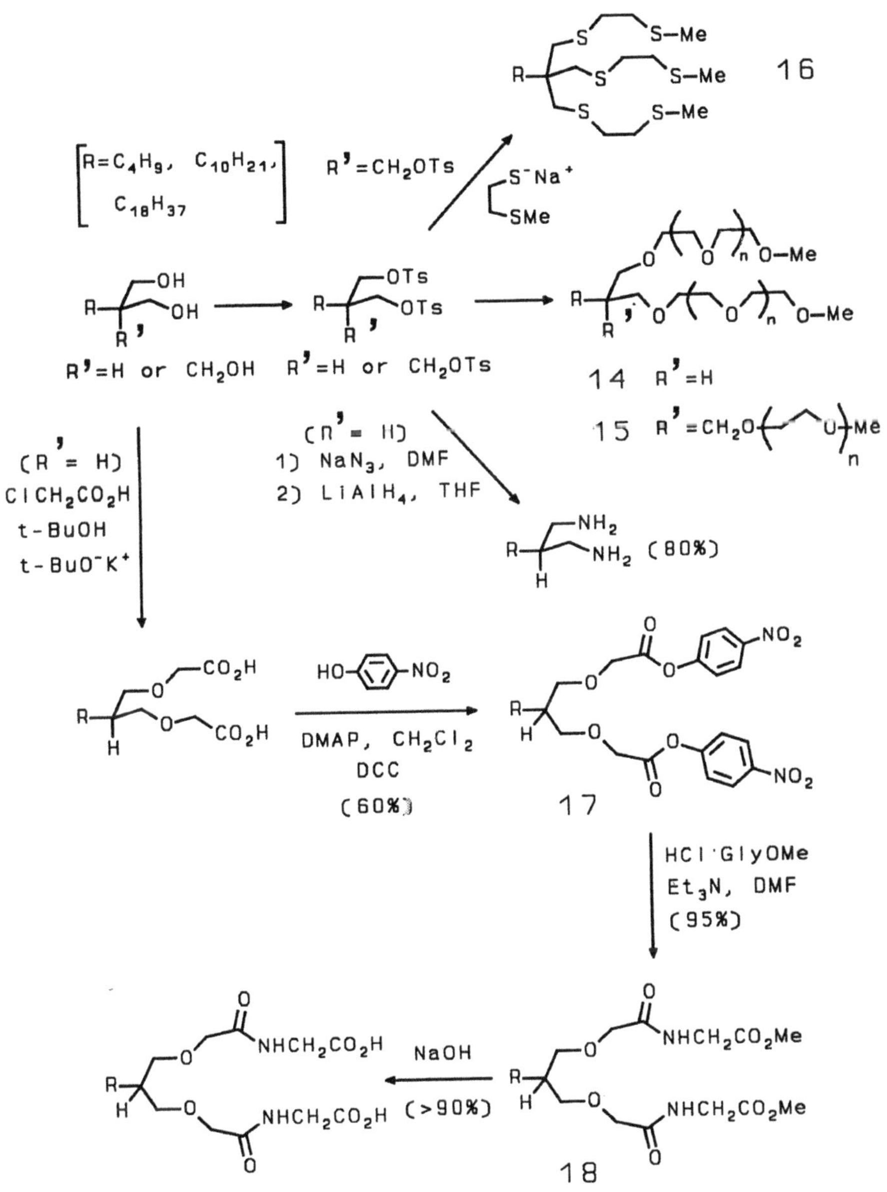

SCHEME 2-6 Activated ester methodology used for the construction of bis(glycine methyl ester) **18**.

methanolysis. Such by-products of nitroalkanes are also isolated when single-electron transfer (SET) conditions are employed.[36] Facile elimination of HNO_2 from angular tertiary nitroalkanes upon contact with silica gel or neutral alumina at ambient temperature[37] or just by distillation in vacuo[38] affords access to novel trisubstituted polyfunctional olefins.

POLYAZIDO CASCADES

The synthesis of a cascade molecule incorporating energetic azide moieties via other pentaerythritol derivatives has been reported.[39] Reaction of *bis*(3,3-azidomethyl)oxetane with HBr or HNO_2, followed by treatment with NaN_3, yielded pentaerythritol triazide **19** (Scheme 2-7), which upon addition to 1,3,5-benzenetricarbonyl trichloride provided the three-directional nonaazide **20**. Understandably, most of the starting materials and products were reported to be moderately to considerably sensitive explosives!

The recent successful preparation[40] of related mixed donor building blocks, (2- and 3-aminopropyl)diphenylphosphines from 1-azido-2-propanol and 1-azido-3-chloropropane, respectively, afforded entrance to a new series of polyhetero cascade molecules.

PHOSPHINE BASED CASCADES

Polyphosphorus ligands (e.g., **21**; Figure 2-3) have been prepared[41] with the intention of stabilizing transition metals in low oxidation states. The resulting polyphosphino ligands were subsequently subjected to elemental sulfur to prepare the P=S derivatives; 9 of the possible 19 products with one or more P=S bonds have been characterized. An alternate route to substituted tertiary phosphines such as *tris*(tetrathiafulvalenyl)-phosphine[42] is demonstrated by the treatment of tetrathiafulvalene with lithium diisopropylamide, followed by addition of PBr_3 at low temperatures under an inert atmosphere.

FUNCTIONALIZED CROWNS

Lehn's Chundles

Another interesting example of polyfunctionality, manifested in a multiarmed cascade molecule, has been reported by Jullien and Lehn.[43] The octacarboxylate of polyester **22** was prepared as a mimic of transmembrane molecular channels such as the molecular tubes or channels[44,45]

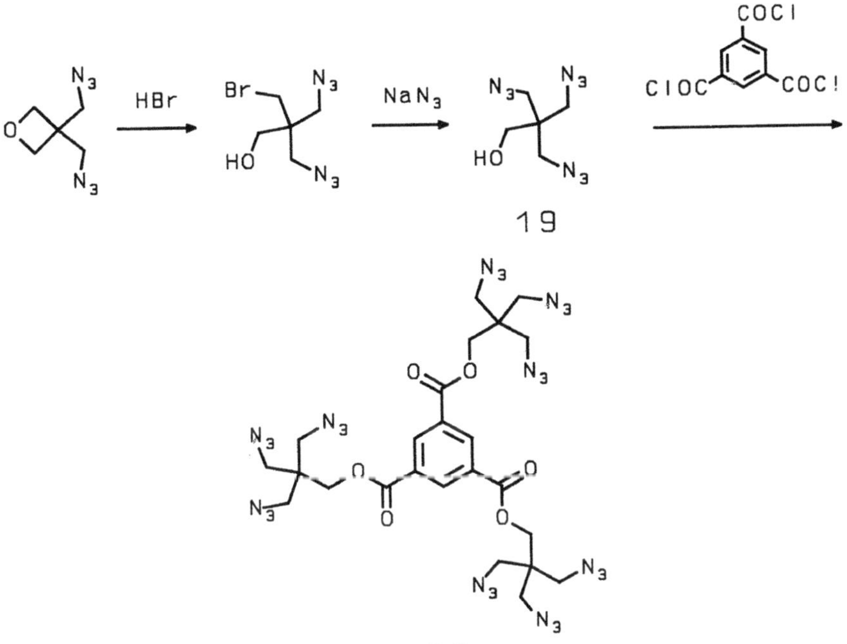

HBr NaN3 19 20

SCHEME 2-7 Construction of a nonaazide.

formed by a stacked array of macrocycles. Preparation of **22** involved the grafting of preformed arms (**23**) onto a four-directional core consisting of a tetraacyl substituted 18-crown-6 macrocyclic polyether (**24**). The arms were constructed (Scheme 2-8) from the outside (in a convergent route) via condensation of an *N*-benzyloxyformylated amino acid with an aminopolyether. Reductive transformation of the urethane group to

FIGURE 2-3 A phosphine-based cascade molecule.

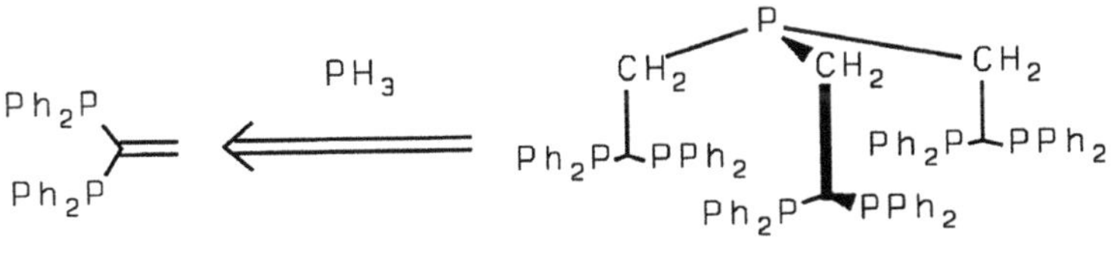

PH_3 21

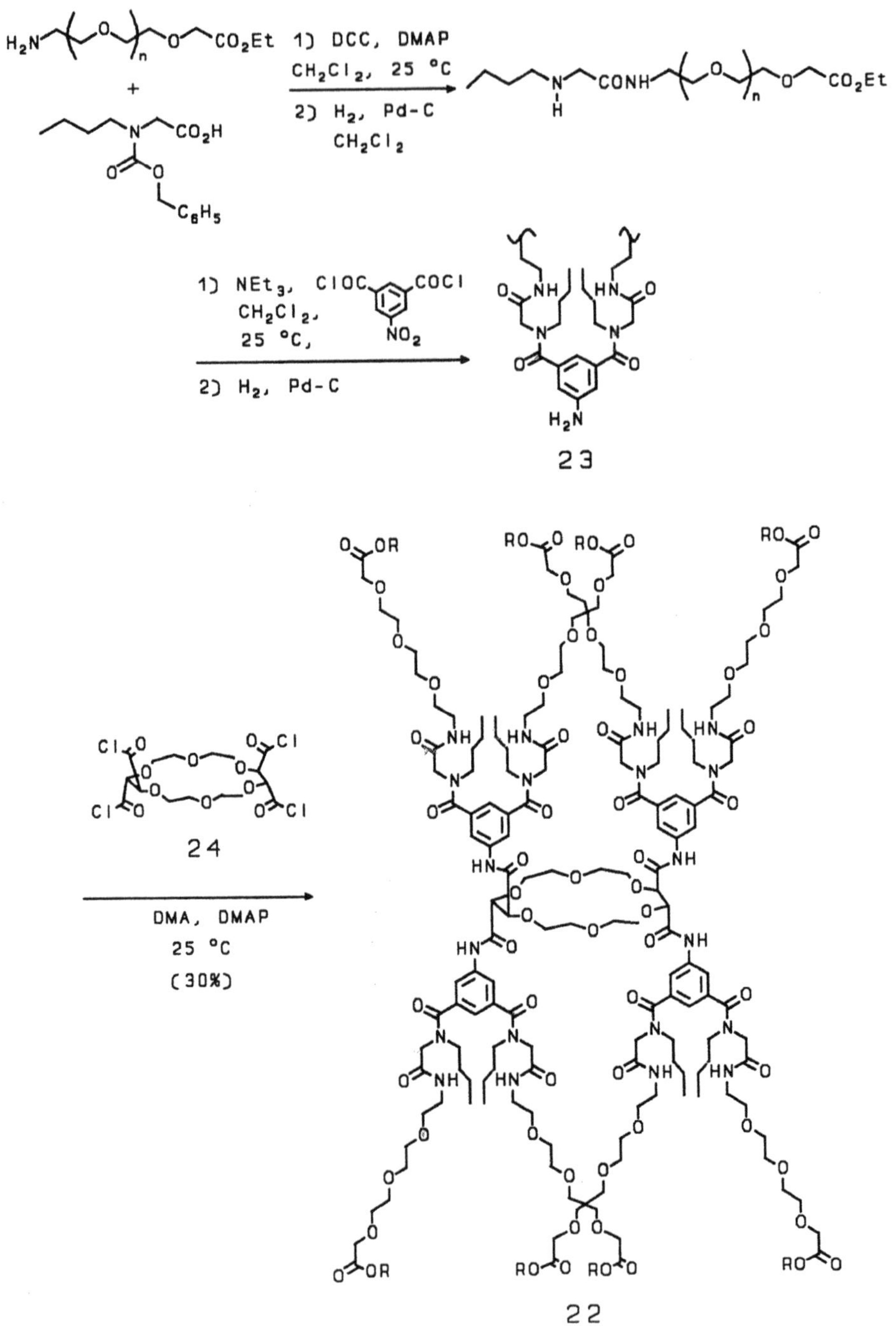

SCHEME 2-8 The "chundle" approach to the preparation of macrocycle-based molecular channels.

the secondary amine and subsequent reaction with 1,3-di(chloroacyl)-5-nitrobenzene followed by catalytic reduction of the nitro group afforded the diester building block containing a three-directional aromatic center of branching. The arms, **23**, were attached to the core, **24**, via amino acylation to give the desired cascade molecule **22**, which is termed a *chundle* molecule because it contains a *ch*annel formed by b*undles* of polyether arms; it is also of nanometer size and represents one of the first examples of such a mesomolecule.

Azamacromolecule Based

Use of a six-directional core[45] (**25**) with appropriate lypophilic appendages afforded macromolecule **26**, which stacks in a supramolecular arrangement possessing tubular (columnar) mesophases (Scheme 2-9). Markovitsi et al.[46] investigated the related belted semiplanar eight-directional cascade molecule, *octakis*(octadecyloxymethyl)phthalocyanine **27**, that formed similar columnar mesophases (Figure 2-4). Laser excitation in homogeneous benzene solutions and in pure columnar mesophases of **27** was used to prove mesophase transitions. Phase transition temperatures were determined: crystal → mesophase, 62 °C; mesophase → isotropic liquid, 193 °C.

CASCADE HYDROCARBON

A synthetic approach to a single-tier, all-hydrocarbon, four-directional cascade molecule by Mendenhall et al.[47] involved alkylation of a trialkyl substituted carbon to generate a quaternary carbon via an *in situ* generation of a trialkyl aluminum reagent prepared from *t*-BuCH=CH_2 and *i*-Bu_3Al. The resulting tetraalkylmethane derivative **28** was obtained in a 70% yield (Scheme 2-10).

CASCADE RADICALS AND ANIONS

Another example of cascade methodology was recently provided by Rajca and coworkers.[48-52] Research with the goal of preparing ferromagnetic materials that do not contain transition metals pointed to the synthesis of large, high-spin, stable organic molecules (Figure 2-5).[53,54] Thus, monolithiation of 1,3-dibromobenzene, addition of 4,4′-di-*tert*-butylbenzophenone, and subsequent treatment with ethyl chloroformate (Et-OCOCl) yielded the cascade arms. Subjection of the resulting triarylbromide to halogen-metal exchange allowed attachment to the

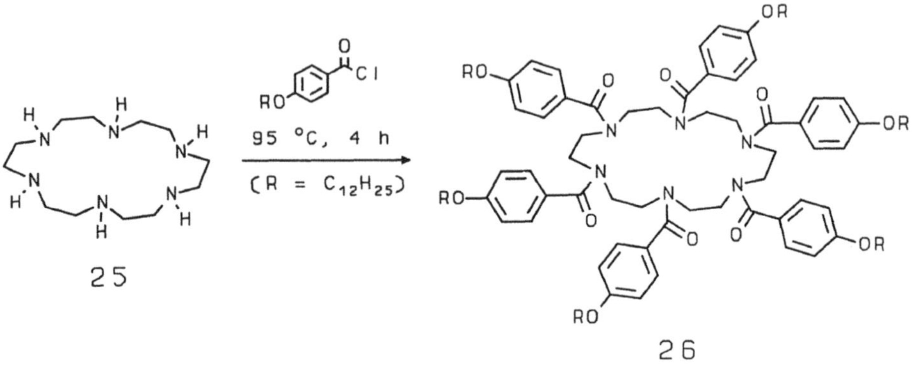

95 °C, 4 h

(R = $C_{12}H_{25}$)

25

26

SCHEME 2-9 Employment of a hexaaza core for the synthesis of cascade molecules that form tubular mesophases.

FIGURE 2-4 Columnar mesophases formed by *octakis*(octadecyl-oxy-methyl)phthalocyanine (**27**).

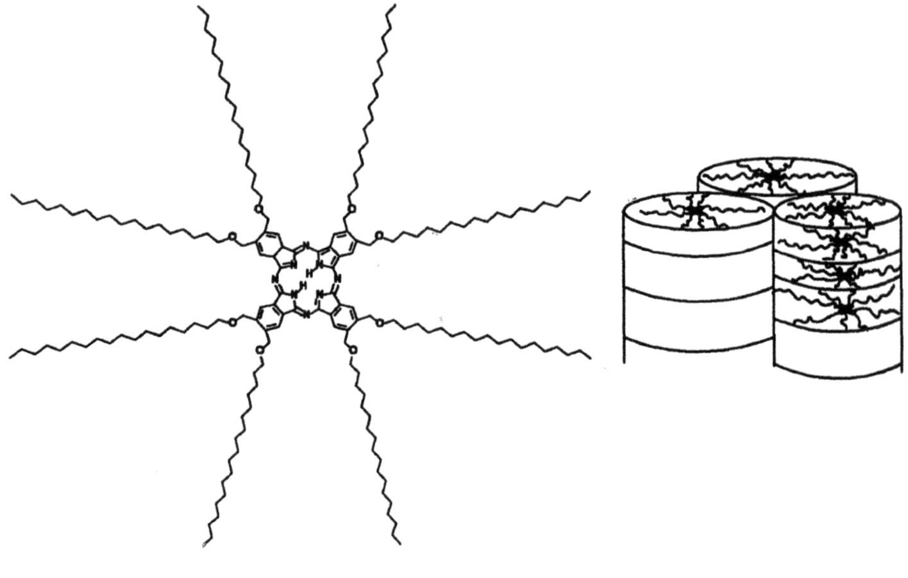

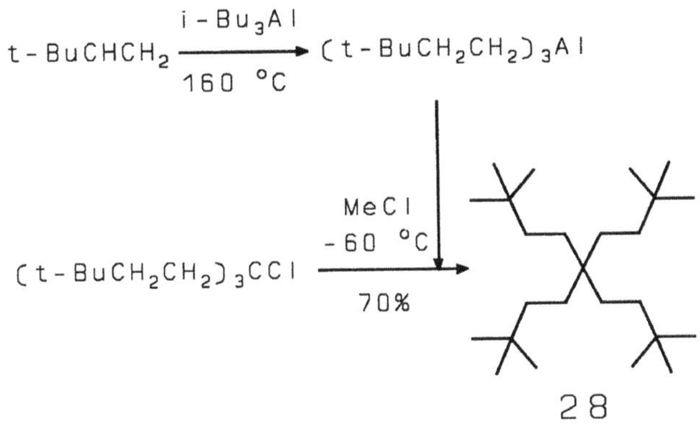

SCHEME 2-10 The use of a trialkyl aluminum building block for the preparation of a hydrocarbon "molecular fractal."

three-directional core, dimethyl carbonate. The resulting oxyanion, which was generated by treatment with Li metal, was trapped with EtOCOCl, yielding the tetraethoxy precursor to tetracarbanion **29**. Oxidation of the polyanion with I_2 gave the corresponding tetraradical **30**.

Kirste, Grimm, and Kurreck[55] also investigated a tetraradical, **32**, except that this polyradical possessed tetrahedral symmetry imparted by the four-directional tetraphenylmethane core, **31**. The unique core was synthesized by Friedel-Crafts alkylation of aniline with triphenylmethanol, diazotization and deamination, bromination, and finally

FIGURE 2-5 Iterative methodology was used to prepare tetraanion **29** and tetraradical **30.**

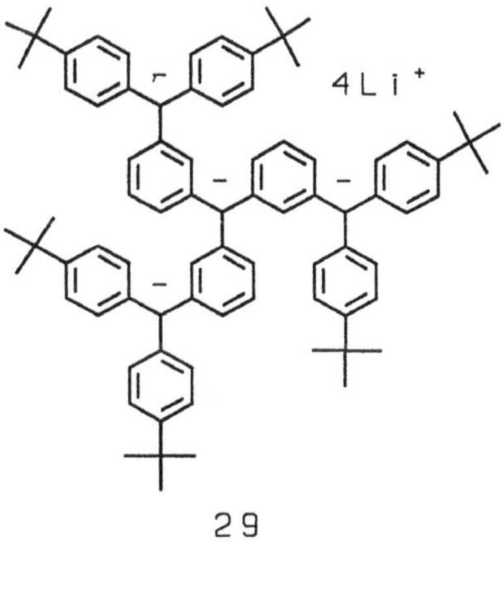

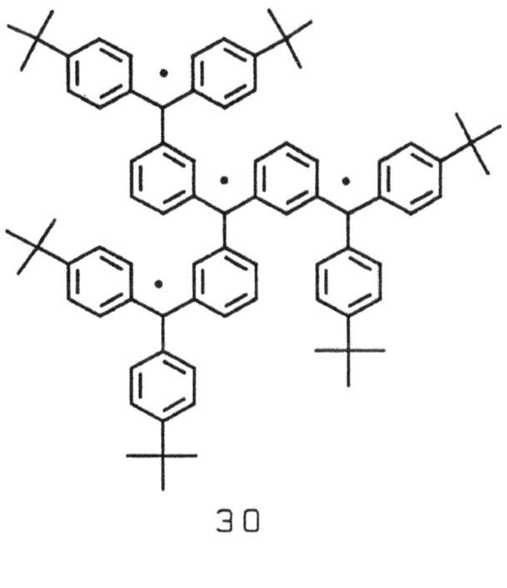

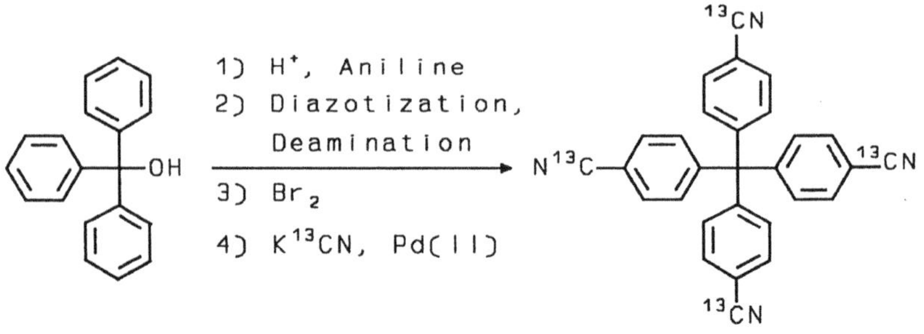
OH
1) H+, Aniline
2) Diazotization, Deamination
3) Br2
4) K13CN, Pd(II)
13CN
N13C
13CN
13CN

31

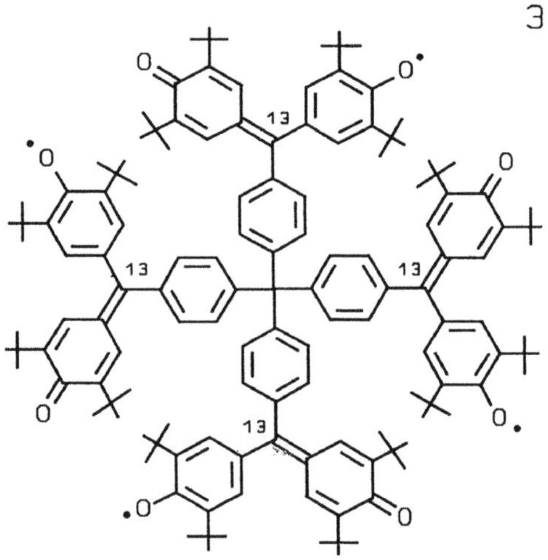
13
13
13
13
O
O•
•O
O
O
O•
•O
O

32

SCHEME 2-11 Use of a tetrahedral, tetraphenyl core for the synthesis of polyradical **32.**

$Pd(OAc)_2$-catalyzed cyanation (Scheme 2-11). Galvinolation was accomplished with (2,6-di-*tert*-butyl-4-lithiophenoxy)-trimethylsilane, and the polyradical was generated via treatment with $K_3[Fe(CN)_6]/KOH$.

POLYFERROCENES

The preparation of a dodecaferrocene, **33,** for the study of multiple redox active centers attached to an organic host molecule (Scheme 2-12) has been reported.° The core tetraferrocene intermediate **34** was synthesized[57,58] via a standard one-step, acid-catalyzed condensation of resor-

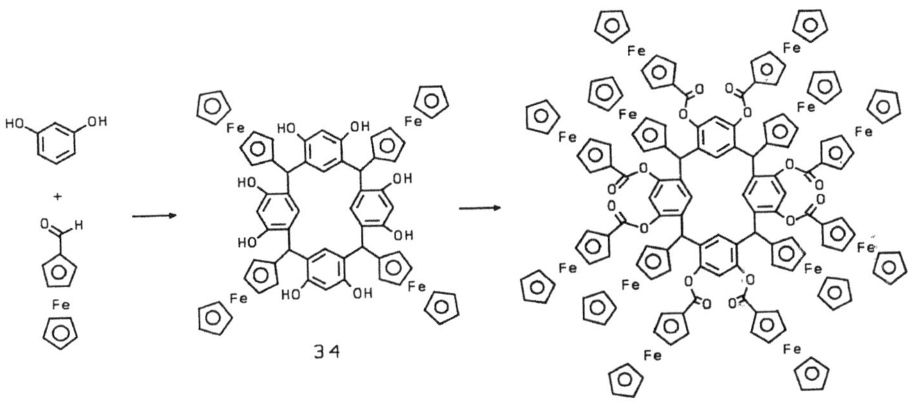

SCHEME 2-12 Preparation of a cascade molecule possessing 12 redox active ferrocene moieties (**33**).

cinol and ferrocene-carboxaldehyde; *O*-acylation with ferrocenecarboxylic acid chloride afforded the dodecabranched product **33.**

FROM MOTHER NATURE

Sweet Cascades

A fascinating use for cascade molecules was provided by Hanessian et al.[59] via the preparation of "megacaloric cluster compounds" (Scheme 2-13). The carbohydrate arms were preconstructed and subsequently attached to four- and five-directional cores. Benzyl 2,3,4-tri-*O*-benzyl-α-D-glucopyranoside was treated with succinic anhydride, in the presence of a catalyst, to yield a protected sugar possessing a terminally functionalized side chain for attachment to a suitable core, such as pentaerythritol and glucose. Traditional coupling with DCC and 4-pyrrolidinopyridine of this building block afforded clusters **35** and **36**, respectively, which upon hydrogenolysis yielded the corresponding completely deprotected polyols.

Polypeptides

Circularly organized trispeptides were reported by Tor et al.[60] The unique organization was attributed to intramolecular H-bonding, based on nuclear magnetic resonance (NMR) and infrared (IR) spectroscopic data, as well as empirical force field calculations. Preparation of trispeptide **37** was accomplished by coupling the core, *tris*-1,3,5-(aminome-

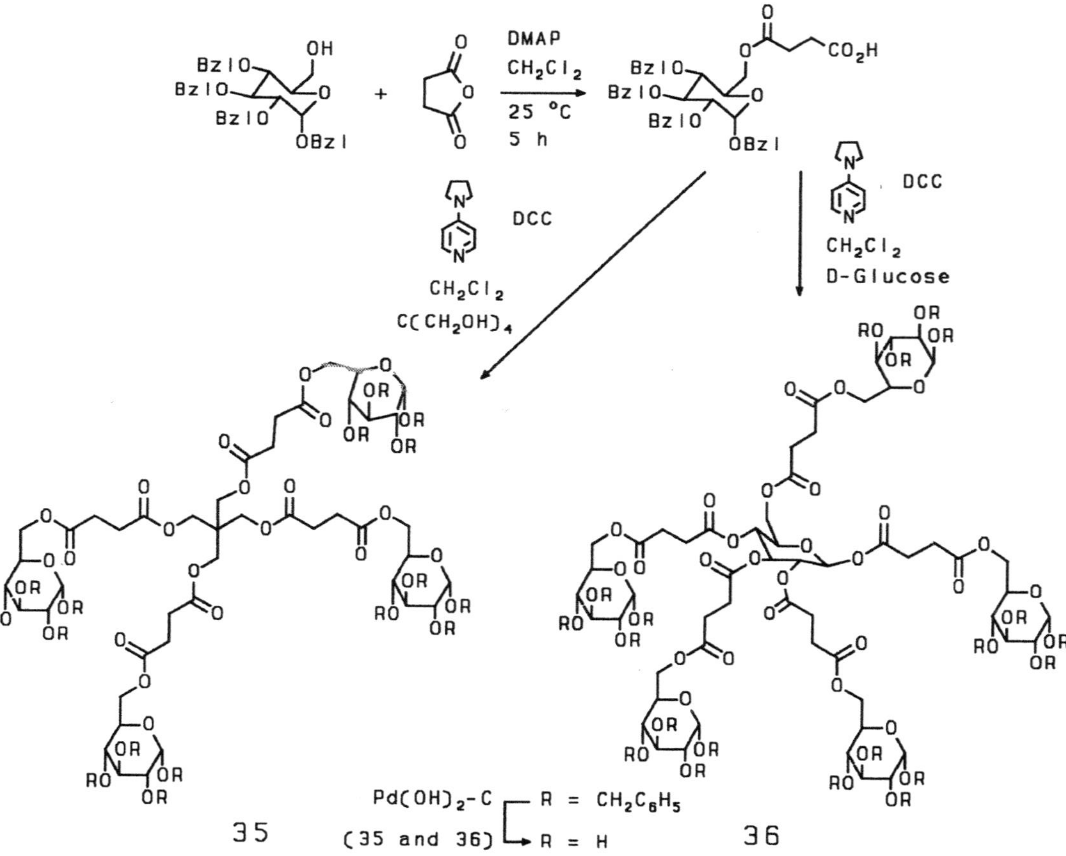

SCHEME 2-13 Preparation of "megacaloric cluster cascade molecules."

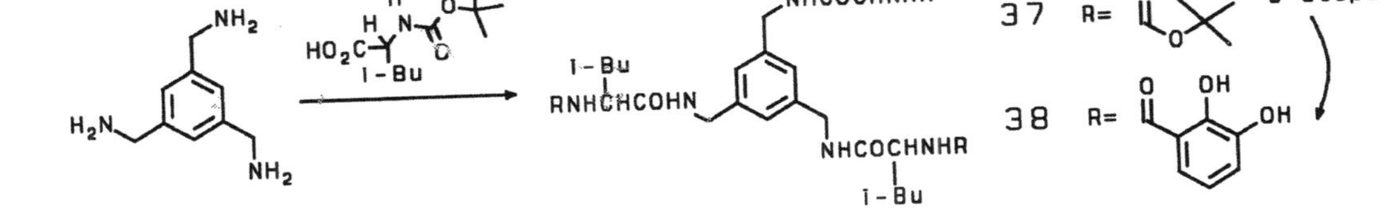

SCHEME 2-14 Construction of an enterobactin siderophore analog (38) starting with a three-directional, triamine core.

thyl)benzene, with *N*-Boc-L-leucine via the *N*-hydroxysuccinimide actived ester (Scheme 2-14). Subsequent deprotection of the terminal amine moiety, reaction with 2,3-bis(benzyloxy)benzoyl chloride, and hydrogenolysis provided enterobactin siderophore analog **38.**[61-63] A series of related trifunctional cores coupled with other amino acids, such as histidine and methionine, has also been reported.[64]

Poly-*N*-Bases

The synthesis of a novel, heteroditopic hexaimidazole ligand (**41**) starting with histamine (Scheme 2-15) has been described.[65] Treatment of histamine with *N,N'*-carbonyldiimidazole, followed by CH_3I and subsequent acid hydrolysis, afforded the *N*-methylhistamine **39.** Protection of the free amine **39** was achieved by silylation with Me_3SiCl. Then lithiation at low temperatures, followed by addition of diethyl carbonate, yielded trisimidazolyl-alcohol **40**, which, with a precursor to **39**, gave rise to the hexadentate ligand **41.** This ligand was reported[65] to be useful

SCHEME 2-15 Preparation of a heteroditopic hexaimidazole ligand (**41**) using diethylcarbonate for core formation.

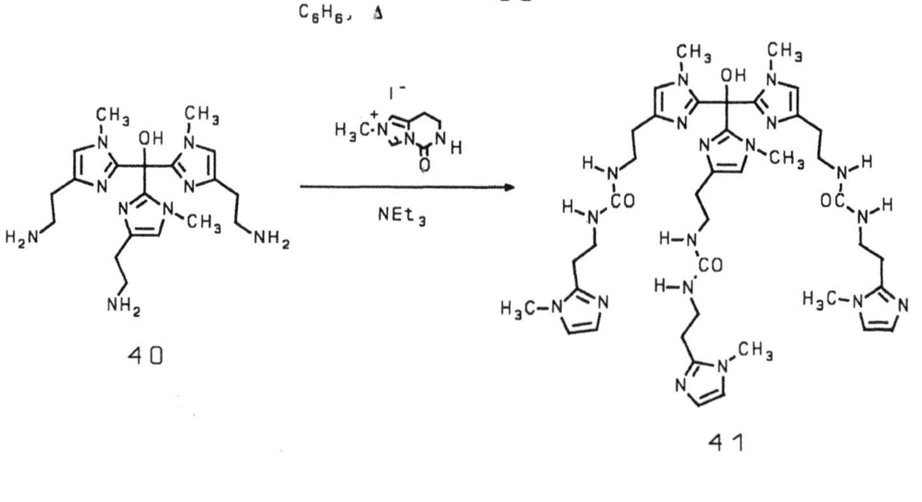

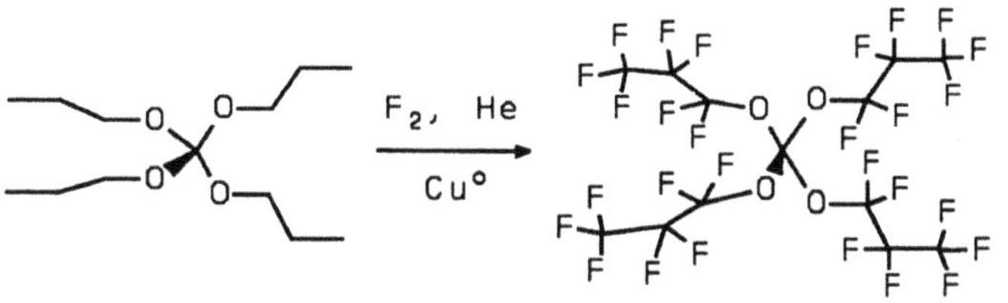

SCHEME 2-16 "Spherical" fluorocarbons prepared via perfluorination of a four-directional tetraether.

for the preparation of dimetallic complexes, as well as for modeling the active site for O_2-transport in hemerythrin.

SLIPPIN' AND SLIDIN'

An interesting example (Scheme 2-16) of "spherical" fluorocarbons was recently provided.[66] Tetra-*n*-propyl orthocarbonate was mixed with NaF, dispersed onto copper turnings, and subjected to "direct fluorination," affording perfluorotetra-*n*-propyl orthocarbonate (**42**). The four-directional, perfluorinated cascade molecule was prepared in conjunction with an investigation into the properties of "spherical" lubricants. It should be noted that this is an excellent one-step procedure to produce fluoro[2.2.2]cryptane[67] and should be readily applicable to many cascade molecules.

FRENCH SANDWICHES

Astruc et al.[68-71] have reported the preparation of fascinating six-directional organoiron complexes that are stable under many diverse reaction conditions. Quantitative yields of hexaalkylated benzene, such as **43**, were obtained via treatment of a hexamethylbenzene cyclopentadienyl iron sandwich complex with allyl bromide and base (Scheme 2-17). The core, hexaalkene **43**, was easily transformed to (1) dodecabromide **44** via Br_2 addition; (2) hexasilane **45** by Pt-mediated silation; (3) hexaol **46** with 9-borabicyclo[3.3.1]nonane (9-BBN) and HO_2^-; and (4) hexaester **47** via acylation of polyol **46**. Decomplexation of the hexa-substituted arene ligands was accomplished by visible-light photolysis

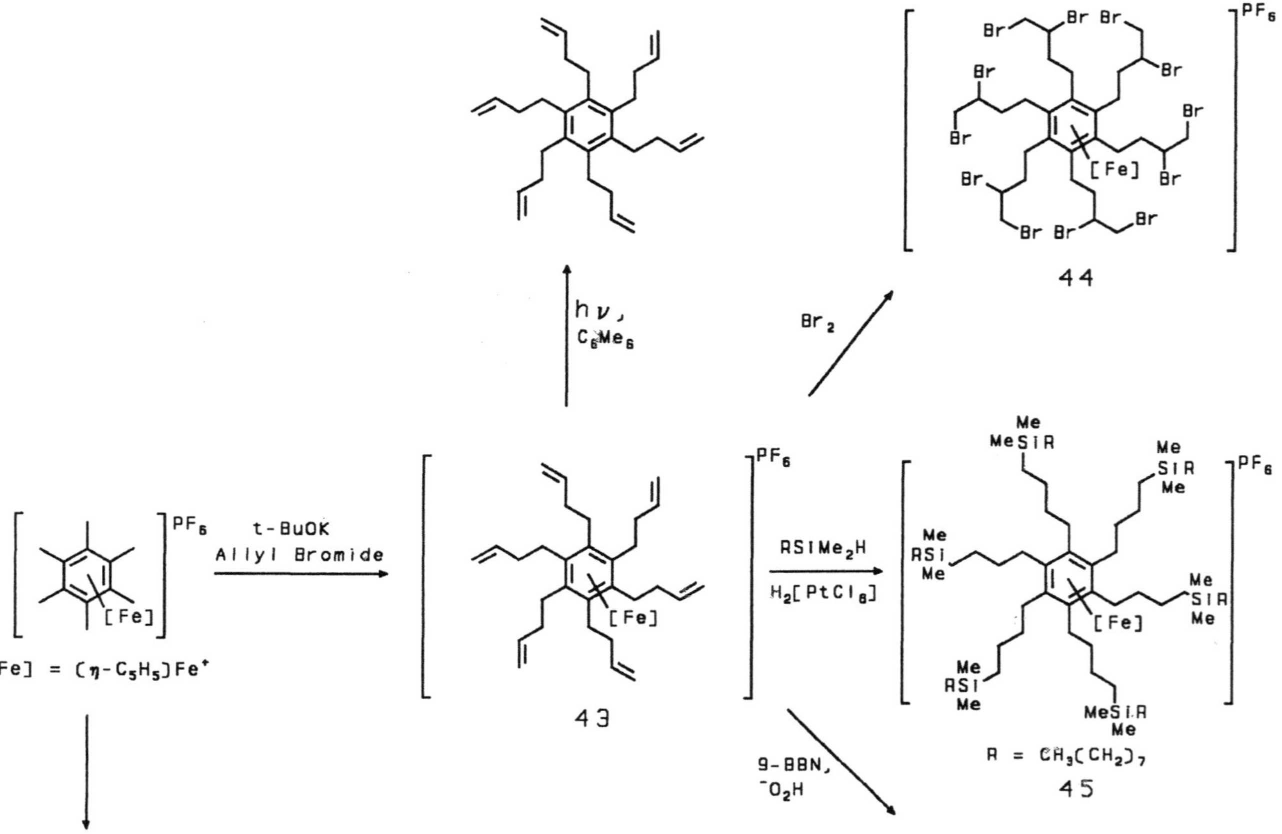

[Fe] = (η-C5H5)Fe+
PF6
t-BuOK
Allyl Bromide
43
hν, C6Me6
Br2
44
RSiMe2H
H2[PtCl6]
9-BBN, -O2H
R = CH3(CH2)7
45

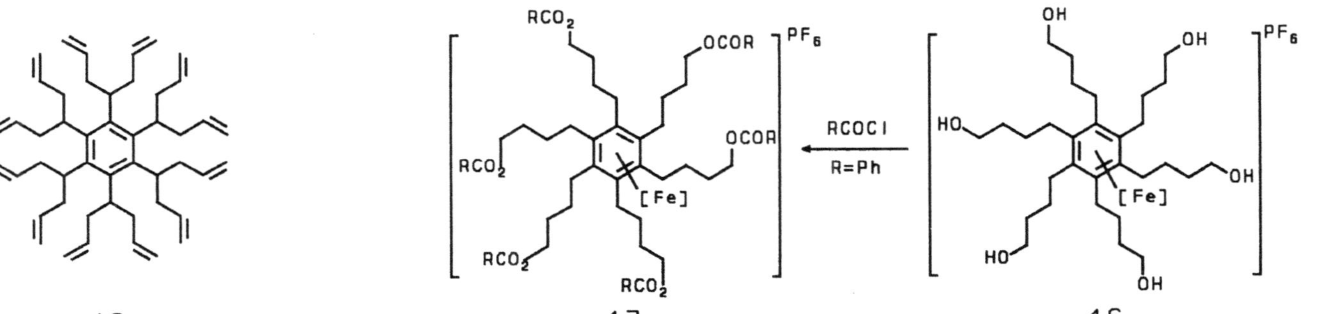

SCHEME 2-17 The preparation of polyfunctional organoiron complexes that offer exciting possibilities as potential cores for cascade polymers.

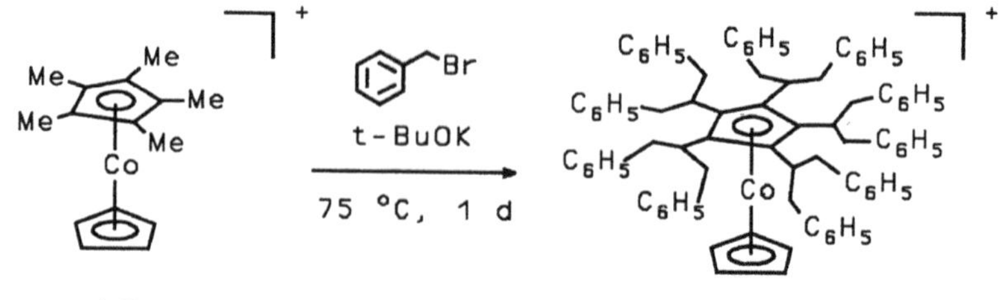

SCHEME 2-18 Synthesis of cobaltocene cascade molecules.

in the presence of hexamethylbenzene, which regenerated the starting material.

With prolonged reaction times, the starting hexamethyl iron complex or **43** with allyl bromide produced double branching and thus simple entrance to cascade polymers ("arboroles" **48**).[72] Moulines et al.[72] have also treated cobaltocene **49** with benzyl bromide to afford **50** in 45% yield (Scheme 2-18), thus demonstrating that polyfunctionalization of π-ligands can be conducted by appropriate activation. This example epitomizes the thesis of this chapter: that cascade molecules (first-tier growth) can be readily transformed into large cascade polymers by the addition of other appropriately chosen polyfunctional building blocks.

BONN'S BONDED BEAUTIES

It is impossible to discuss cascade molecules without recognizing Vögtle's notable contributions to this area. He not only coined the orig-

SCHEME 2-19 Pd-catayzed trimerization afforded the dodecabromide **51.**

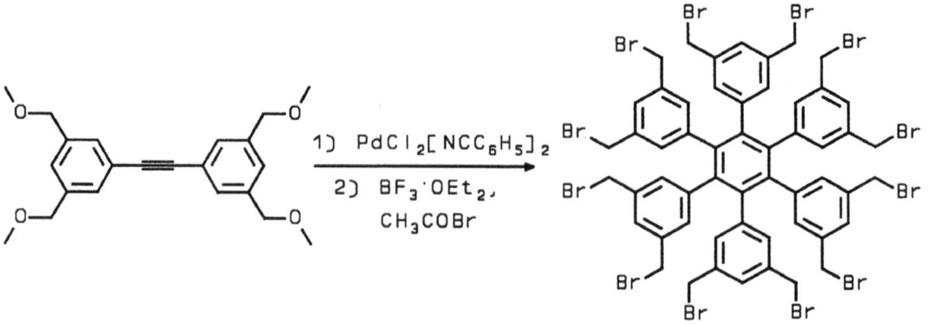

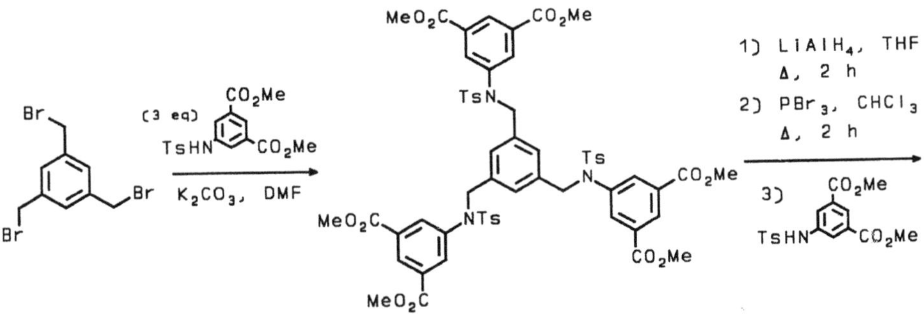

SCHEME 2-20 Synthesis of a dodececaester cascade polymer (**53**) that could serve as a core for continued modification.

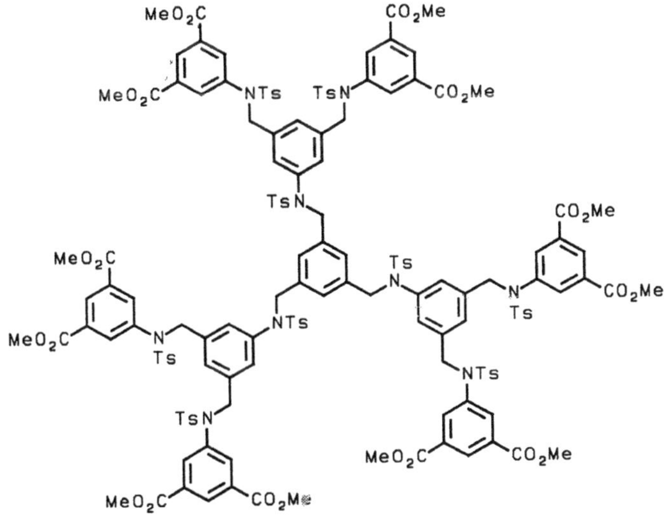

inal term *cascade,*[73] but his research group has prepared the most novel and diverse polyfunctional mesomolecules yet created.[74] It should be noted that most of their mesomolecules will make fascinating cores for cascade polymers.

Continuing investigations into the preparation of polyfunctional molecules, Duchêne and Vögtle[75] synthesized an interesting "dodeca" core (Scheme 2-19). Palladium-catalyzed trimerization of a bis[(dimethoxymethyl)-phenyl]alkyne, followed by benzylic bromination, afforded paddle wheel-like cascade molecule **51.** Upon

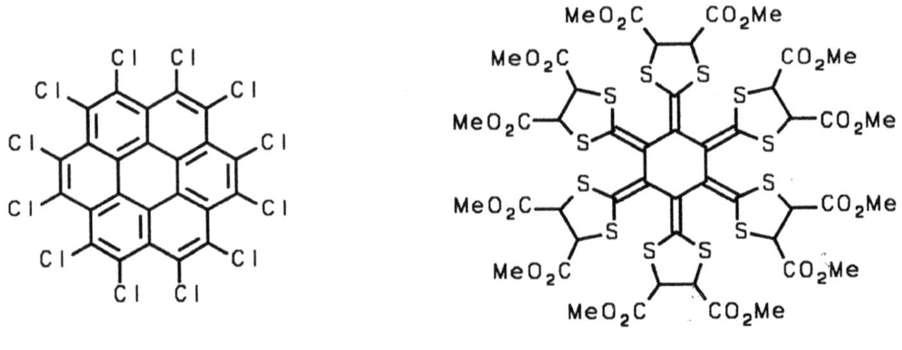

54

55

FIGURE 2-6 Dodeca-directional, polychloro and polyester cores (**54** and **55**, respectively).

crystallization of **51** from cyclohexane, a 2:1 cyclohexane clathrate was formed, which was verified by ^{1}H NMR and elemental analysis. This dodeca-directional cascade core provides interesting possibilities to dodeca directionality in subsequent tier construction with numerous building blocks, as noted by the successful conversion of **51** to the corresponding dodecathiomethyl ether.

Recently, simple cascade construction (Scheme 2-20) has also been demonstrated[76] by Vögtle and coworkers, in which 1,3,5-tris(bromomethyl)-benzene with the building block derived from the *N*-tosylate of dimethyl 5-aminoisophthalate afforded hexaester **52**. Re-

SCHEME 2-21 Construction of a "dendralene" hexaester, **57**, beginning with homologated tetrathiafulvalene **56**.

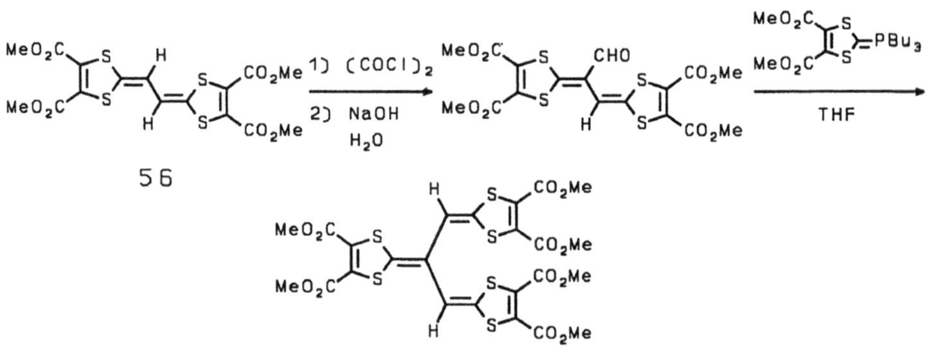

56

1) $(COCl)_2$
2) NaOH H_2O

THF

57

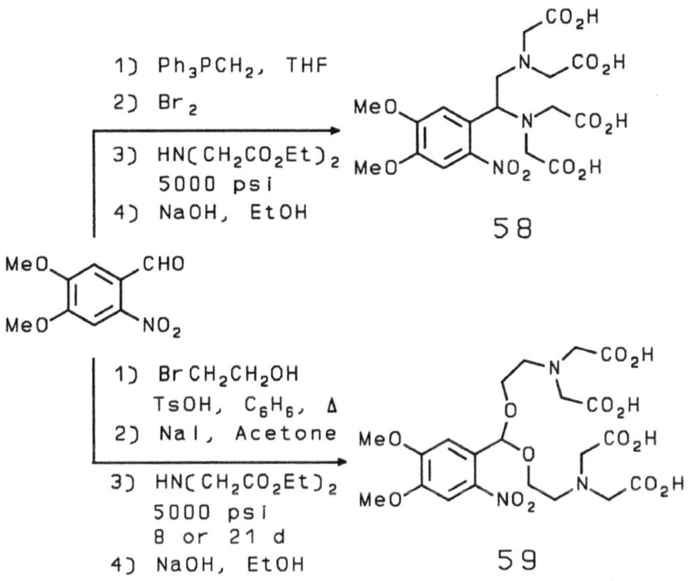

1) Ph_3PCH_2, THF
2) Br_2
3) $HN(CH_2CO_2Et)_2$
5000 psi
4) NaOH, EtOH

58

1) $BrCH_2CH_2OH$
TsOH, C_6H_6, Δ
2) NaI, Acetone
3) $HN(CH_2CO_2Et)_2$
5000 psi
8 or 21 d
4) NaOH, EtOH

59

SCHEME 2-22 High pressure (5000 psi) was employed for the preparation of tetra-acetic acid derivative **59.**

duction with $LiAlH_4$ afforded the hexaol, which was subsequently treated with PBr_3 to give the corresponding hexabromide; this, when reacted with the same arylamide building block, generated the dodecaester **53.** The third generation was similarly constructed, giving rise to the nanoscale cascade polymer with 24 surface ester moieties. Although "dense packing" (surface congestion inhibiting subsequent complete conversion to the next tier) will be a factor for continued growth, other building blocks were envisioned.

FIGURE 2-7 EDTA (**60**) and "structurally constrained" EDTA (**61**).

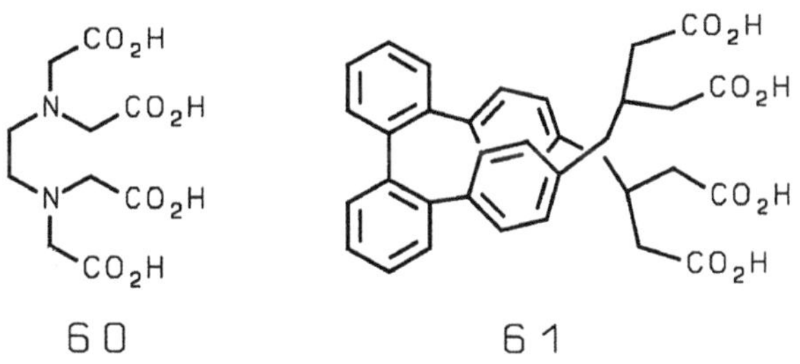

60

61

DENDRALENES AND CORONENES

Another potential dodeca-directional core (Figure 2-6) was reported by MacNicol et al.,[77,78] whereby coronene was perchlorinated with a mixture of $AlCl_3$, SCl_2, and SO_2Cl_2. The ability to perform *complete* substitutions on perchlorinated coronene **54** in excellent yields was demonstrated by treatment with sodium arenethiolate in 1,3-dimethylimidazolidin-2 as solvent. A striking example of a dodeca-directional monomer, *hexakis*[bis(4,5-dicarbomethoxy)-1,3-dithiole-2-ylidene]cyclohexane (**55**), which was prepared by Ni^0/Zn-Cu-mediated coupling of 2,2′-(1,2-dibromoethanediylidene)*bis*(4,5-dicarbomethoxy-1,3-dithiole), has recently been reported.[79] X-ray determinations revealed that the cyclohexane moiety of the [6]radialene possessed a twisted-boat conformation promoted by steric crowding (Figure 2-6). Related "dendralenes" such as **56** (Scheme 2-21) have been synthesized,[80] employing a Vilsmeyer-type reaction using oxalyl chloride on **57**, followed by a Wittig reaction with 1,3-dithiol-2-ylidene tri-*n*-butylphosphoranes; repetition of this sequence affords homologues possessing useful polyfunctional, linear cascade cores.

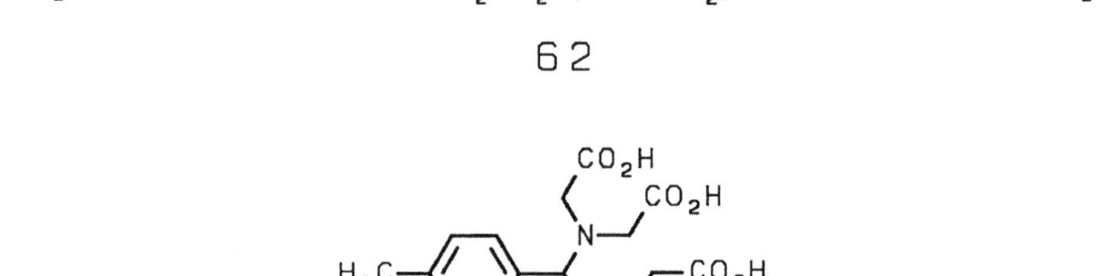

A = $CNOCH_2CO_2H$, $CHNH_2$, CONH

62

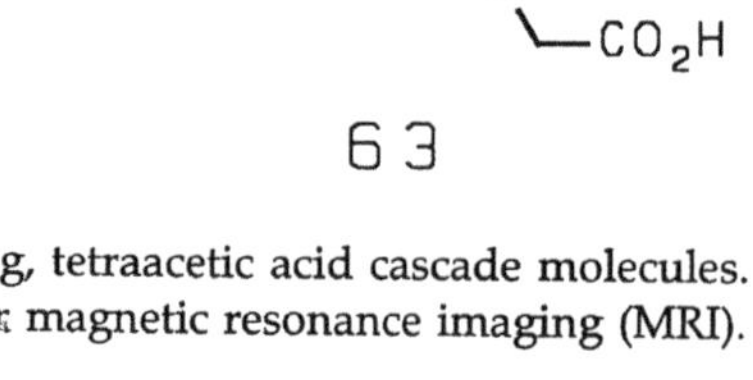

63

FIGURE 2-8 Chelating, tetraacetic acid cascade molecules. Polyacid **63** was prepared to complex gadolinium for magnetic resonance imaging (MRI).

VINAIGRETTE SAUCE

Ellis-Davies and Kaplan[81] reported the synthesis of chelating cascade molecules designed for the photochemical release of divalent cations such as Mg^{2+} and Ca^{2+} (Scheme 2-22). A key step in the preparation of the polyfunctional 2-nitro-4,5-dimethoxyphenyl analogs, **58** and **59**, involved the reaction of iminodiacetic acid diethyl ester with primary alkyl iodide or bromide moieties at elevated pressures for extended periods of time.

Ethylenediamine tetraacetic acid (EDTA) has been a traditional ligand for the complexation of diverse metal ions. When there is an analogy between the stereochemical disposition of the donor functions, such as **60** and **61** (Figure 2-7), Leppkes and Vögtle[82] have suggested a structural "donor stereology." This spatial relationship needs to be expanded from cascade molecules to the larger macromolecular systems. Numerous modifications have been developed over the past decade, such as the polyacid complexing agents, **62**, that incorporated pyridine moieties, which were reported by Takalo, Hänninen, and Kankare[83] (Figure 2-8). Preparation of these complexing agents involved palladium/copper(l) coupling reactions between monosubstituted acetylenes and diiodoarenes. Related endeavors by McLaren, Baker, and Wilson[84] were aimed at the preparation of polyacid ligands (**63**) for gadolinium that would serve as MRI-enhancing agents. The biodistribution of a ^{153}Gd(III)·**63** complex in mice was reported.

AMINES AND PYRIDINES

Karlin and Gultneh[85] have emphasized the interactions of oxygen with discrete soluble copper coordination complexes. These studies have led them to the preparation[86] of the hexapyridyltetraamine **64** to gain a better understanding of multinuclear copper clusters in connection with four-electron reduction of dioxygen to water in multicopper oxidases.[87] This cascade ligand **64** was synthesized (Scheme 2-23) via acid-catalyzed addition of *tris*(2-aminoethyl)amine (tren) to 2-vinylpyridine and was then transformed to a trinuclear copper complex whose structure was confirmed by X-ray analysis. Additional studies of pyridyl-terminated cascade molecules[88-90] resulted in the preparation of tetrapyridyl ligand **65**, which was able to coordinate two Cu(I) ions. This bis(copper) complex exhibited reactivity with dioxygen and was investigated as a copper monooxygenase mimic. A convenient one-step synthesis of a related series of *N,N,N,',N'-tetrakis*(2-pyridylmethyl)-α,ω-alkanediamines (**66**) has recently been described,[91] utilizing tetraalkylammonium chloride as a phase transfer catalyst (Scheme 2-24). The related polyethylenepolyamines,[92] such as **67**, were prepared when linear polyamines were condensed with bis(2-chloroethyl)amine hydrochloride.

BOLAAMPHIPHILES

"Bolaform amphiphiles" (**68**) were prepared by Fyles et al.[93] for the purpose of mimicking the pore-forming antibiotic amphotericin. The two-directional, hydroxyl-terminated amphiphile was synthesized via coupling of two macrocyclic tetraesters with 1,3-bis(thiomethyl)benzene, affording a dimer diene, which was then reacted with sodium β-D-glucose-1-mercaptide (Figure 2-9). Principal spectral features were assigned; however, due to the polyethylene oxide and polymethylene units, the assignment of all resonances associated with each isomer was precluded. Vesicles were prepared using the amphiphiles, and the transport of Na^+ and K^+ cations across the bilayer membranes was determined.

SCHEME 2-23 Preparation of copper-coodinating cascade ligands for the purpose of investigating Cu-oxygenase activity.

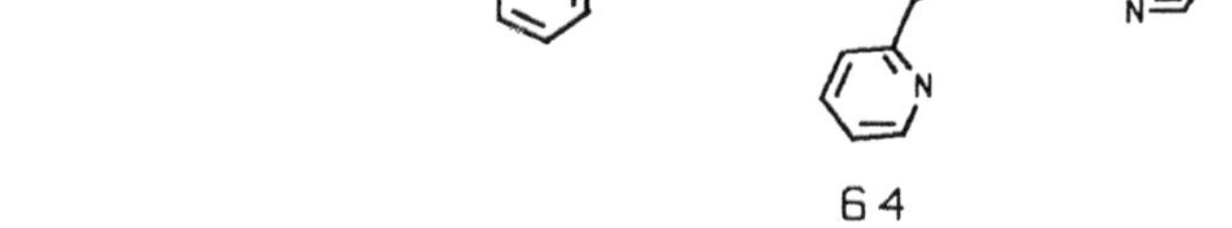

$N(CH_2CH_2NH_2)_3$ → 64

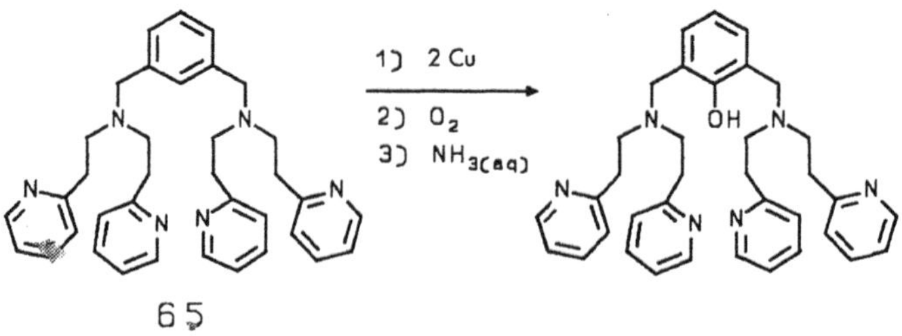

65 → 1) 2 Cu; 2) O_2; 3) $NH_{3(aq)}$

FUNCTIONALLY DIFFERENTIATED BRICKS

Sluka et al.[94] introduced a method whereby cascades **69** and **70** were employed as site-specific, affinity-cleaving agents for DNA (Scheme 2-25). Synthesis of the tribenzyl ester of EDTA (**69**) was achieved by tris-*N*-alkylation of a mono-*N*-alkylated ethylenediamine with benzyl bromoacetate; the tricyclohexyl analog (**70**) was prepared (21%) by tris-esterification of EDTA.

CARBOSILANES

The potential to create diverse polyfunctional molecules has also been demonstrated by Uchida et al.,[95] employing siloxane building blocks for the construction of a series of silicone-based dendrimers. They used silicon chemistry (alkenylation-hydrosilylation) for the preparation of carbosilane cascade molecules and dendrimers.[96,97] The iterative transformations proceeded in nearly quantitative yields and afforded macromolecules possessing up to 972 termini.[96]

SCHEME 2-24 Preparation of a tetrapyridyl cascade chelate (**66**) and a polyethyleneamine ligand (**67**).

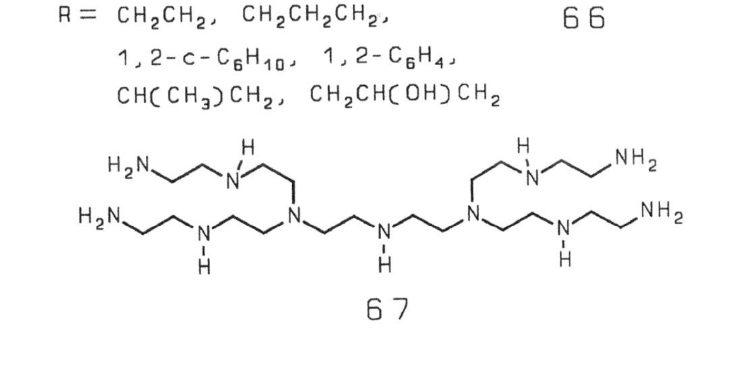

$H_2N-(R)-NH_2$
5N NaOH
$CetMe_3NCl$
H_2O, 20 °C, 24 h
42–86%

66

R = CH_2CH_2, $CH_2CH_2CH_2$,
1,2-c-C_6H_{10}, 1,2-C_6H_4,
$CH(CH_3)CH_2$, $CH_2CH(OH)CH_2$

67

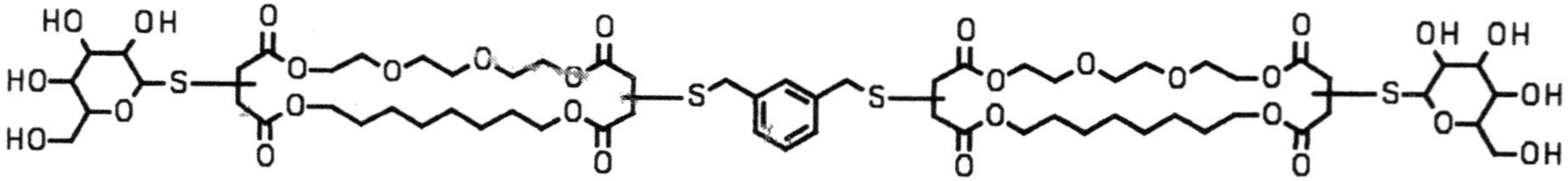

68

FIGURE 2-9 A polyfunctional "bolaform amphiphile" (**68**) synthesized to mimic the pore-forming antibiotic amphotericin.

ENTRANCE—INORGANIC COMPLEXATION

The introduction of inorganic chemistry expands the horizons of cascade molecules (Figure 2-10). Recently, Constable et al.[98-100] incorporated the terpyridine moiety, as denoted by **71**; Beer et al.[101] used the metal atom as the core (e.g., **72**); and Balzani et al.[102,103] used the metal atom as a bridging linkage (see Chapter 3 in this volume). Each group crossed the traditional organic framework construction with the introduction of inorganic functionality to a pivotal component within the mesomolecule. Newkome et al.[104] further melded the simple building blocks[12,105] with Constable's terpyridine[100] moiety to generate the dodecaterpyridine surface **73** at the second tier. Addition of other polyfunctionalized terpyridine building blocks (e.g., **74**) with a Ru(III) site permitted the construction of the next tier (**75**) via a ruthenium bis-terpyridine connector. Such framework design (Scheme 2-26) introduces specifically located metal atoms within the molecular lattice.

MIDSCRIPT

This chapter has afforded a partial overview of cascade molecules that can be readily used in or transformed into diverse homogeneous or heterogeneous cascade polymers.

SCHEME 2-25 Cascade molecules **69** and **70** were prepared to function as site-specific, affinity-cleaving agents for DNA.

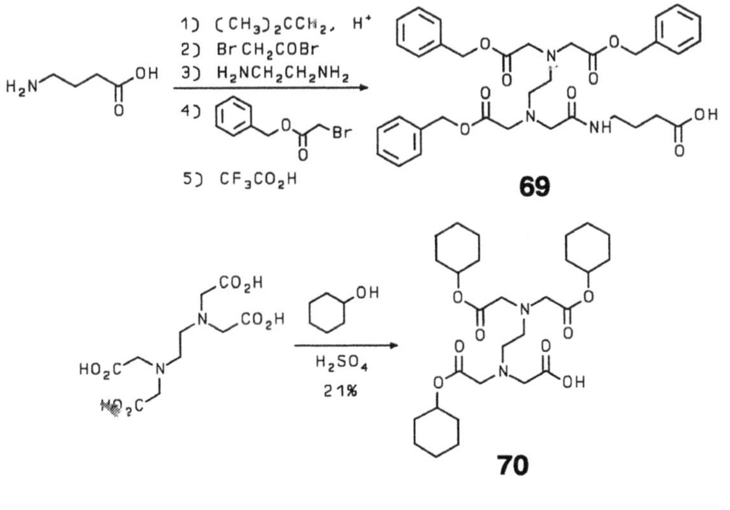

1) $(CH_3)_2CCH_2$, H^+
2) $BrCH_2COBr$
3) $H_2NCH_2CH_2NH_2$
4) [benzyl bromoacetate]
5) CF_3CO_2H

69

H_2SO_4
21%

70

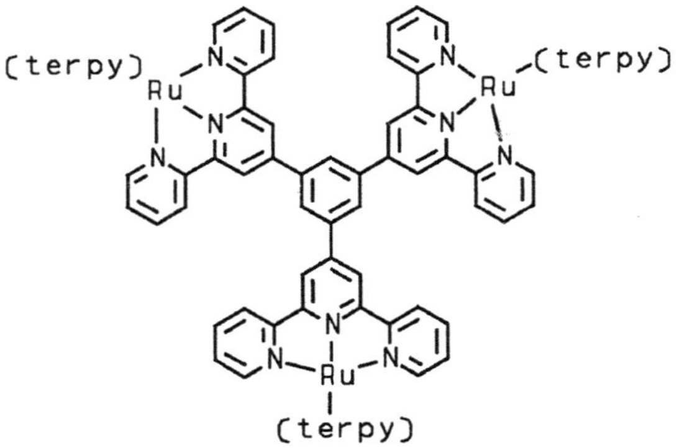

(terpy)=2,2′:6′,2″-terpyridine

71

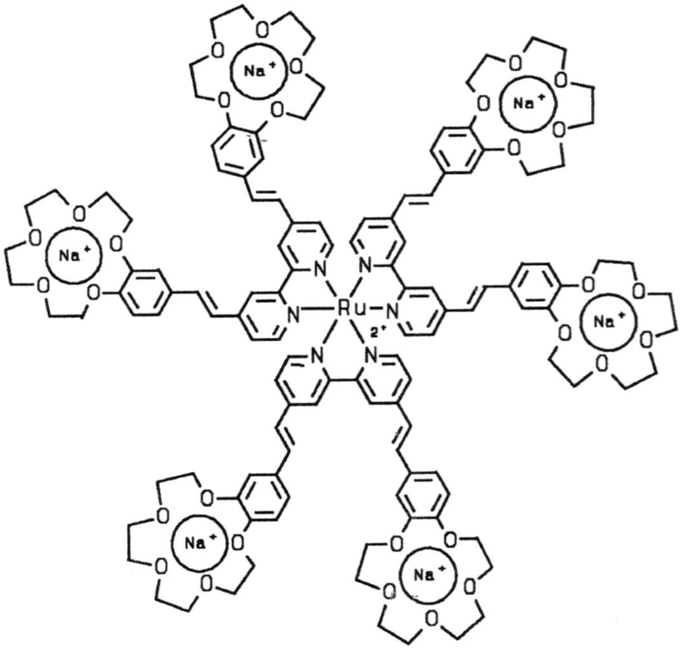

72

FIGURE 2-10 The incorporation of metals for connectivity in cascade molecules.

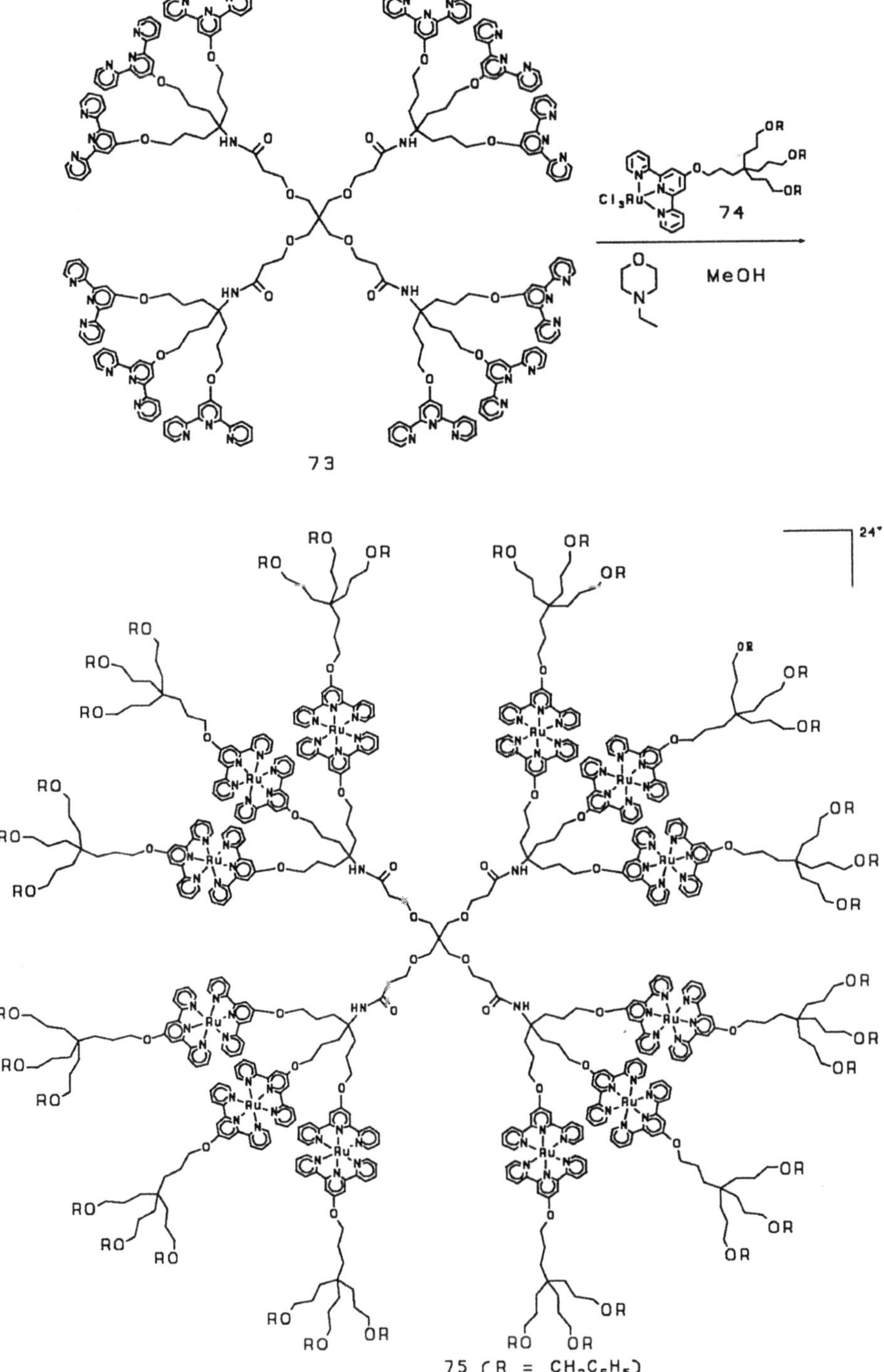

SCHEME 2-26 Preparation of a Ru-terpyridyl cascade (75).

Acknowledgment

Acknowledgment is made to the National Science Foundation and the Petroleum Research Fund, administered by the American Chemical Society, for partial support of this work. Special thanks go to the numerous collaborators who have contributed their time and energy to expand these molecular horizons.

REFERENCES

1. F.M. Menger, *Top Curr. Chem.*, **36,** 1 (1986).
2. I. Amoto, *Science News*, **138,** 297 (1990).
3. Y.-X. Chen, *Youji Huaxue*, **10,** 289 (1990).
4. H. Bernhard, M. Burges, W. Jaworek, and F. Vögtle, *Angew. Chem. Int. Ed. Engl.*, **104,** 1609 (1992); *Angew. Chem.*, **31,** 1571 (1992).
5. K. Krohn, *Organic Synthesis Highlights*, J. Mulzer, Ed. (VCH, New York, 1991, p 378).
6. S. Shinkai, *Kagaku Dojin*, **42,** 74 (1987).
7. F. Vögtle, *Supramolecular Chemistry* (John Wiley, Chichester, 1991).
8. J.-M. Lehn, *Frontiers in Supramolecular Organic Chemistry and Photochemistry*, H.J. Schneider and H. Dürr, Eds. (VCH, Weinheim, 1991), pp 1–28.
9. *Supramolecular Chemistry*, V. Balzani and L. DeCola, Eds. (Kluwer Academic Press, Dordrecht, 1992).
10. G.R. Newkome, Z.-Q. Yao, G.R. Baker, and V.K. Gupta, *J. Org. Chem.*, **50,** 2003 (1985).
11. J. March, *Advanced Organic Chemistry*, 4th ed. (Wiley, New York, 1992), pp 275–286.
12. G.R. Newkome, C.N. Moorefield, and K.J. Theriot, *J. Org. Chem.*, **53,** 5552 (1988).
13. G.R. Newkome and G.R. Baker, *Org. Prep. Proc. Int.*, **18,** 117 (1986).
14. J. Skarżewski, *Synthesis*, 1125 (1990).
15. G.R. Newkome, G.R. Baker, J.K. Young, and J.G. Traynham, *J. Polym. Sci., Part A: Polym. Chem.*, **31,** 641 (1993).
16. G.R. Newkome, G.R. Baker, M.J. Saunders, P.S. Russo, V.K. Gupta, Z.-Q. Yao, J.E. Miller, and K. Bouillon, *J. Chem. Soc., Chem. Commun.*, 752 (1986).
17. G.R. Newkome, G.R. Baker, S. Arai, M.J. Saunders, P.S. Russo, K.J. Theriot, C.N. Moorefield, J.E. Miller, T.R. Lieux, M.E. Murray, B. Phillips, and L. Pascal, *J Am. Chem. Soc.*, **112,** 8458 (1990).

18. G.R. Newkome, C.N. Moorefield, G.R. Baker, R.K. Behera, G.H. Escamilla, and M.J. Saunders, *Angew. Chem. Int. Ed. Engl.*, **31,** 917 (1992).
19. H.R. Byers and K. Fujiwara, *J. Cell Biol.*, **93,** 804 (1982).
20. L.G. Tilney and S. Inoué, *J. Cell Biol.*, **93,** 820 (1982).
21. M.S. Mooseker, *Annu. Rev. Cell Biol.*, **1,** 209 (1985).
22. L.G. Tilney, *Modern Cell Biology, Vol. 2: Spatial Organization of Eukaryotic Cells,* J.R. McIntosh and B.H. Satir, Eds. (Liss, New York, 1983), pp 163–199.
23. A.F. Bochkov, B.E. Kalganov, and V.N. Chernetskii, *Izv. Akad. Nauk. SSSR, Ser. Khim.*, **10,** 2394 (1989).
24. A. McAuley, S. Subramanian, and T.W. Whitcombe, *J. Chem. Soc., Chem. Commun.*, 539 (1987).
25. A. McAuley, S. Subramanian, and W.T. Whitcombe, *Can. J. Chem.*, **67,** 1650 (1989).
26. A. McAuley, K. Beveridge, S. Subramanian, and T.W. Whitcombe, *Can. J. Chem.*, **67,** 1657 (1989).
27. G.R. Newkome, S. Arai, F.R. Fronczek, C.N. Moorefield, X. Lin, and C.D. Weis, *J. Org. Chem.*, **58,** 898 (1993).
28. A.T. Phillips, *Aust. J. Chem.*, **21,** 2301 (1968).
29. J. Skarżewski and J. Młochowski, *Tetrahedron,* **39,** 309 (1983).
30. S.R. Cooper, *Pure & Appl. Chem.*, **62,** 1123 (1990).
31. C.M. Thorne, *Thesis* (Somerville College, Oxford, 1986).
32. R.J. Geue and G.H. Searle, *Aust. J. Chem.*, **36,** 927 (1983).
33. J. Ehrler and D. Seebach, *Liebigs Ann. Chem.*, 379 (1990).
34. J. Skarżewski and J. Młochowski, *J. Chem. Res.* (*S*), 64 (1988).
35. S. Latour and J.D. Wuest, *Synthesis,* 742 (1987).
36. N. Ono, H. Miyake, R. Tanikaya, and A. Kaji, *J. Org. Chem.*, **47,** 5017 (1982).
37. A.P. Marchand and S.C. Suri, *J. Org. Chem.*, **49,** 670 (1984).
38. C.D. Weis and G.R. Newkome, *J. Org. Chem.*, **55,** 5901 (1990).
39. E.R. Wilson and M.B. Frankel, *J. Org. Chem.*, **50,** 3211 (1985).
40. S.-T. Liu and C.-Y. Liu, *J. Org. Chem.*, **57,** 6079 (1992).
41. J.L. Bookman and W. McFarlane, *Polyhedron,* **29,** 129 (1988).
42. M. Fourmigué and P. Batail, *J. Chem. Soc., Chem. Commun.*, 1370 (1991).
43. L. Jullien and J.-M. Lehn, *Tetrahedron Lett.*, **29,** 3803 (1988).
44. U.F. Kragton, M.F.M. Roks, and R.J.M. Nolte, *J. Chem. Soc., Chem. Commun.*, 1275 (1985).
45. J.-M. Lehn, J. Malthête, and A.-M. Levelut, *J. Chem. Soc., Chem. Commun.*, 1794 (1985).
46. D. Markovitsi, T.-H. Tran-Thi, V. Briois, J. Simon, and K. Ohta, *J. Am. Chem. Soc.*, **110,** 2001 (1988).

47. G.D. Mendenhall, S.X. Liang, and E.H.-T. Chen, *J. Org. Chem.*, **55,** 3697 (1990).
48. A. Rajca, *J. Am. Chem. Soc.*, **112,** 5889 (1990).
49. A. Rajca, *J. Am. Chem. Soc.*, **112,** 5890 (1990).
50. S. Utamapanya and A. Rajca, *J. Am. Chem. Soc.*, **113,** 9242 (1991).
51. A. Rajca and S. Utamapanya, *J. Am. Chem. Soc.*, **114,** 1760 (1992).
52. A. Rajca, S. Utamapanya, and S. Thayumanavan, *J. Am. Chem. Soc.*, **114,** 1884 (1992).
53. K. Nakatani, J.Y. Carriat, Y. Journaoux, O. Kohn, F. Lloret, J.P. Renard, Y. Pei, J. Sletten, and M. Verdaguer, *J. Am. Chem. Soc.*, **111,** 5739 (1989).
54. A. Caneschi, D. Gatteschi, R. Sessoli, and P. Rey, *Acc. Chem. Res.*, **22,** 392 (1989).
55. B. Kirste, M. Grimm, and H. Kurreck, *J. Am. Chem. Soc.*, **111,** 108 (1989).
56. P.D. Beer and E.L. Tite, *Tetrahedron Lett.*, **29,** 2349 (1988).
57. P.D. Beer, M.G.B. Drew, A. Ibbotson, and E.L. Tite, *J. Chem. Soc., Chem. Commun.*, 1498 (1988).
58. P.D. Beer, O. Kocian, R.J. Mortimer, and C. Ridgway, *J. Chem. Soc., Chem. Commun.*, 1460 (1991).
59. S. Hanessian, C. Hoornaert, A.G. Vernet, and A.M. Nadzan, *Carbohydr. Res.*, **137,** C14 (1985).
60. Y. Tor, J. Libman, A. Shanzer, C.E. Felder, and S. Lifson, *J. Chem. Soc., Chem. Commun.*, 749 (1987).
61. Y. Tor, J. Libman, A. Shanzer, and S. Lifson, *J. Am. Chem. Soc.*, **109,** 6517 (1987).
62. Y. Tor, J. Libman, A. Shanzer, C.E. Felder, and S. Lifson, *J. Am. Chem. Soc.*, **114,** 6653 (1992).
63. Y. Tor, J. Libman, A. Shanzer, C.E. Felder, and S. Lifson, *J. Am. Chem. Soc.*, **114,** 6653 (1992).
64. D. Goldfarb, J.-M. Fauth, Y. Tor, and A. Shanzer, *J. Am. Chem. Soc.*, **113,** 1941 (1991).
65. P.G. Potvin and M.H. Wong, *J. Chem. Soc., Chem. Commun.*, 672 (1987).
66. W.-H. Lin, W.D. Clark, and R.J. Lagow, *J. Org. Chem.*, **54,** 1990 (1989).
67. W.D. Clark, T.Y. Lin, S.D. Maleknia, and R.J. Lagow, *J. Org. Chem.*, **55,** 5933 (1990).
68. D. Astruc, *Topics Curr. Chem.*, **160,** 47 (1991).
69. D. Astruc, M.-H. Desbois, B. Gloaguen, M. Moulines, and J.-R. Hamon, *Organic Synthesis via Organometallics*, K.H. Dötz and H.W. Hoffman, Eds. (Vieweg, Braunschweig, 1991), pp. 63–75.
70. F. Moulines and D. Astruc, *Angew. Chem. Int. Ed. Engl.*, **27,** 1347 (1988).
71. F. Moulines, L. Djakovitch, J.-L. Fillaut, and D. Astruc, *Synlett*, 57 (1992).
72. F. Moulines, B. Gloaguen, and D. Astruc, *Angew. Chem. Int. Ed. Engl.*, **31,** 458 (1992).

73. E. Buhlier, W. Wehner, and F. Vögtle, *Synthesis,* 155 (1978).
74. F. Vögtle, *Cyclophan-Chemie,* (B.G. Teubner, Stuttgart, 1990).
75. K.-H. Duchêne and F. Vögtle, *Synthesis,* 659 (1986).
76. H.-B. Mekelburger, K. Rissanen, and F. Vögtle, *Chem. Ber.,* **126,** 1161 (1993).
77. T. Baird, J.H. Gall, D.D. MacNicol, P.R. Mallinson, and C.R. Michie, *J. Chem. Soc., Chem. Commun.,* 1471 (1988).
78. D.D. MacNicol, *Inclusion Compounds,* Vol. 2 (Academic Press, London, 1984), pp. 123–168.
79. T. Sugimoto, Y. Misaki, T. Kajita, Z.-I. Yoshida, Y. Kai, and N. Kasai, *J. Am. Chem. Soc.,* **109,** 4106 (1987).
80. Y. Misaki, Y. Matsumura, T. Sugimoto, and Z.-I. Yoshida, *Tetrahedron Lett.,* **30,** 5289 (1989).
81. G.C.R. Ellis-Davies and J.H. Kaplan, *J. Org. Chem.,* **53,** 1966 (1988).
82. R. Leppkes and F. Vögtle, *Angew. Chem. Int. Ed. Engl.,* **20,** 396 (1981).
83. H. Takalo, E. Hänninen, and J. Kankare, *Acta Chem. Scand.,* **B42,** 662 (1988).
84. A. McLaren, R.J. Baker, and J.G. Wilson, *Aust. J. Chem.,* **40,** 449 (1987).
85. K.D. Karlin and Y. Gultneh, *Prog. Inorg. Chem.,* **35,** 219 (1987).
86. K.D. Karlin, R.W. Cruse, Y. Gultneh, A. Farooq, J.C. Hayes, and J. Zubieta, *J. Am. Chem. Soc.,* **109,** 2668 (1987).
87. A. Meserschmitt, A. Rossi, R. Ladenstein, R. Huber, M. Bolognesi, G. Gatti, A. Marchesini, R. Pettruzelli, and A. Finazzi-Agrò, *J. Mol. Biol.,* **206,** 513 (1989).
88. K.D. Karlin, R.W. Cruse, and Y. Gultneh, *J. Chem. Soc., Chem. Commun.,* 599 (1987).
89. K.D. Karlin, R.W. Cruse, M.S. Haka, Y. Gultneh, and B.I. Cohen, *Inorg. Chim. Acta,* **125,** L43 (1986).
90. K.D. Karlin, Q. Gan, A. Farooq, S. Liu, and J. Zubieta, *Inorg. Chem. Acta,* **165,** 37 (1989).
91. M. Sato, Y. Mori, and T. Iida, *Synthesis,* 539 (1992).
92. M.M. Zgoda and S. Petri, *Pol. Patent,* PL 127 609 (1985).
93. T.M. Fyles, K.C. Kaye, T.D. James, and D.W.M. Smiley, *Tetrahedron Lett.,* **31,** 1233 (1990).
94. J.P. Sluka, J.H. Griffin, D.P. Mack, and P.B. Dervan, *J. Am. Chem. Soc.,* **112,** 6369 (1990).
95. H. Uchida, Y. Kabe, K. Yoshino, A. Kawamata, T. Tsumuraya, and S. Masamune, *J. Am. Chem. Soc.,* **112,** 7077 (1990).
96. A.W. van der Made and P.W.N.M. van Leeuwen, *J. Chem. Soc., Chem. Commun.,* 1400 (1992).
97. J. Roovers, P.M. Toporowski, and L.-L. Zhou, *Polymer Preprints,* **33,** 182 (1992).

98. E.C. Constable, *Tetrahedron,* **48,** 10013 (1992).
99. E.C. Constable, A.M.W.C. Thompson, and D.A. Tocher, *Supramolecular Chemistry,* V. Balzani and L. DeCola, Eds. (Kluwer, Dordrecht, 1992), pp. 219–233.
100. E.C. Constable, A.M.W.C. Thompson, D.A. Tocher, and M.A.M. Daniels, *New J. Chem.,* **16,** 855 (1992).
101. P.D. Beer, J.W. Wheeler, and C. Moore, *Supramolecular Chemistry,* V. Balzani and L. DeCola, eds. (Kluwer, Dordrecht, 1992), pp. 105–118.
102. G. Denti, S. Campagna, S. Serroni, M. Ciano, and V. Balzani, *J. Am. Chem. Soc.,* **114,** 2944 (1992).
103. L. DeCola, F. Barigelletti, V. Balzani, P.V. Belser, A. Zelewsky, C. Seel, M. Frank, and F. Vögtle, *Supramolecular Chemistry,* V. Balzani and L. DeCola, Eds. (Kluwer, Dordrecht, 1992), pp. 157–180.
104. G.R. Newkome, F. Cardullo, E.C. Constable, C.N. Moorefield, and A.M.W.C. Thompson, *J. Chem. Soc., Chem. Commun.,* 925 (1993).
105. G.R. Newkome and X. Lin, *Macromolecules,* **24,** 1443 (1991).

3

Dendritic Polynuclear Metal Complexes with Made-to-Order Luminescent and Redox Properties

GIANFRANCO DENTI, SEBASTIANO CAMPAGNA, AND VINCENZO BALZANI

INTRODUCTION

The current strategy to pursue regarding the miniaturization of electronic components is based on a "large-downward" approach.[1] Further miniaturization, however, as well as the development of a variety of molecular machines,[2] will require a "small-upward" approach based on the assembly of functionally integrated building blocks into structurally organized arrays.[3-17] Supramolecular chemistry has shown that molecular components can indeed be assembled to give species of nanometric dimensions. This bridge from molecules to materials is most promising and will certainly be the object of extensive research in the near future.

The simplest method for the synthesis of well-defined, ultralarge, supramolecular species is the *cascade* approach,[18] which can lead to tree-like structures (arborols or dendrimers[19,20]). Of particular interest are tree-like structures that incorporate in their building blocks specific pieces of information, such as electronic excited states and redox levels at accessible energy values. To pursue this aim, we have designed a synthetic strategy to build up arborols based on luminescent and redox-active transition metal complexes, whereby desired metals and ligands can be placed in specific sites of the supramolecular structure. Species made of 2, 3, 4, 6, 7, 10, 13, and 22 metal-based units have already been prepared, and larger species can in principle be obtained. This strategy opens the way to the synthesis of arborols of nanometric dimension capable of harvesting visible light and exhibiting made-to-order patterns for energy and electron migration.

BUILDING BLOCKS

In order to build up supramolecular species capable of exhibiting redox activity and light-induced functions, appropriate building blocks (components) have to be used. In the last 20 years, extensive investigations carried out on the photochemical and electrochemical properties of transition metal compounds have shown that Ru(II) and Os(II) complexes of aromatic *N*-heterocycles, such as $Ru(bpy)_3^{2+}$ and $Os(bpy)_3^{2+}$ (bpy = 2,2′-bipyridine), exhibit a unique combination of chemical stability, redox properties, excited-state reactivity, luminescence, and excited-state lifetime.[21-27] Furthermore, all of these properties can be tuned within rather broad ranges by changing ligands or ligand substituents. Several hundred of these complexes have been synthesized and characterized, and some of them have been used as sensitizers for the interconversion of light and chemical energy, as well as for other purposes.

It can be shown that the ground state, the low-energy excited states, and the redox forms of these complexes can be described in a sufficiently approximate way by localized molecular orbital configurations.[28] With such an assumption, the various electronic transitions are classified as metal-centered (MC), ligand centered (LC), and charge transfer (CT: either metal to ligand, MLCT, or ligand to metal, LMCT), and the oxidation and reduction processes can be considered as MC and LC, respectively. The properties of $Ru(bpy)_3^{2+}$, the prototype of this class of compounds, are illustrated in Figure 3-1, and those of a few compounds of this large family are summarized in Table 3-1.[21] These complexes exhibit very intense LC bands in the ultraviolet (UV) region and intense MLCT bands in the visible region. Regardless of the excitation wavelength, the originally populated, excited states undergo fast, radiationless decay to the lowest triplet ^{3}MLCT, which is luminescent both in rigid matrix at 77 K and in fluid solution at room temperature. This is a very useful property because, as we will see later, the occurrence of energy migration can be monitored by luminescence measurements.

BRIDGING LIGANDS

In order to assemble such metal-containing building blocks in a supramolecular array, a variety of bridging ligands (BL) can be employed. We

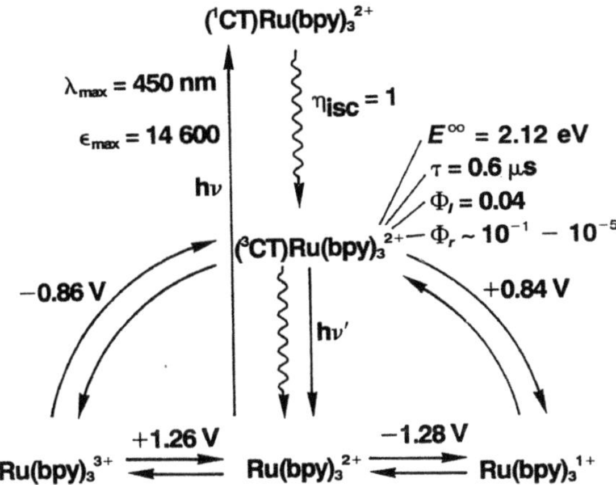

FIGURE 3-1 Schematic representation of some relevant ground- and excited-state properties of $Ru(bpy)_3^{2+}$. ^1CT and ^3CT are the spin-allowed and spin-forbidden MLCT excited states, responsible for the high-intensity absorption band with λ_{max} = 450 nm and the luminescence band with λ_{max} = 615 nm, respectively. The other quantities shown are intersystem crossing efficiency (η_{isc}); energy (E^{oo}) and lifetime (τ) of the ^3CT state; luminescence quantum yield (Φl); and quantum yield for ligand detachment (Φr). The reduction potentials of couples involving the ground and the ^3CT excited states are also indicated.

have concentrated our efforts on the use of the 2,3- and 2,5-bis(2-pyridyl)pyrazine (2,3- and 2,5-dpp) BL shown in Figure 3-2.

Some important properties of the $M(L)_n(BL)_{3-n}^{2+}$ compounds can be summarized as follows (Table 3-2)[24,29-32]: (1) there are intense LC absorption bands in the UV region and moderately intense (ϵ_{max} ~ 1×10^4 M^{-1} cm^{-1}) MLCT bands in the visible region; (2) a relatively long-lived luminescence (10^{-7} to 10^{-8} s at room temperature) is present in the red spectral region; (3) reversible one-electron oxidation of the metal ion takes place in the potential window +0.8/+1.7 V (vs. SCE); (4) reversible one-electron reduction of each ligand takes place in the potential window −0.6/−1.1 V. Important differences relevant to our discussion are as follows: (i) Os(II) complexes are oxidized at potentials considerably less positive than those of Ru(II) complexes; (ii) the MLCT absorption and luminescence bands lie at lower energies for the Os(II) complexes than for the Ru(II) ones; (iii) the energy of the lowest unoccupied molecular orbital (LUMO) of the (monocoordinated) ligands decreases in the series bpy > 2,3-dpp > 2,5-dpp > biq; as a consequence, the lowest (luminescent) ^{3}MLCT level involves the lowest ligand of the above series that is present in the complex; (iv) the electron donor power decreases in the ligand series bpy > biq ≳ 2,3-dpp ~ 2,5-dpp.

TABLE 3-1 Spectroscopic and Electrochemical Properties of Some Ru(II)-diimine Complexes[a]

Complex	Absorption		Emission (293 K)		E_{oo}	E(ox)	E(red)	*E(ox)	*E(red)
	λ, nm	ε, M^{-1} cm^{-1}	λ, nm	τ, μs	eV	.V	V	V	V
$Ru(bpy)_3^{2+}$	452	13,000	615	1.10	2.13	1.26	−1.35	−0.87	0.78
$Ru(bpy)_2(DM\text{-}bpy)^{2+}$	448	11,500	620	0.74	2.08	1.22	−1.37	−0.86	0.71
$Ru(bpy)(DM\text{-}bpy)_2^{2+}$	453	12,200	625	0.72	2.10	1.18	−1.41	−0.92	0.69
$Ru(DM\text{-}bpy)_3^{2+}$	456	10,800	625	0.25	2.08	1.15	−1.46	−0.93	0.62
$Ru(bpy)_2(DTB\text{-}bpy)^{2+}$	450	14,500	625	1.17	2.10	1.21	−1.37	−0.89	0.73
$Ru(bpy)(DTB\text{-}bpy)_2^{2+}$	454	15,300	630	1.07	2.10	1.16	−1.39	−0.94	0.71
$Ru(DTB\text{-}bpy)_3^{2+}$	456	16,800	625	1.15	2.16	1.11	−1.44	−1.05	0.72
$Ru(bpy)_2(biq)^{2+}$	526	8,710	742	0.27	1.70	1.33	−0.91	−0.37	0.79
$Ru(bpy)(biq)_2^{2+}$	547	8,550	742	0.20	1.69	1.40	−0.82	−0.29	0.87
$Ru(biq)_3^{2+}$	524	9,000	705	[b]	1.73	1.47	−0.90	−0.26	0.83

[a]Data from ref. 21. DM-bpy is 3,3′-dimethyl-2,2′-bipyridine; DTB-bpy is 4,4′-di-*tert*-butyl-2,2′-bipyridine; biq is 2,2′-biquinoline. All the data are in deaerated acetonitrile solution. In electrochemical experiments, Ag/$AgNO_3$ was the reference electrode and NHE (normal hydrogen electrode) the standard electrode.

[b]Not measured.

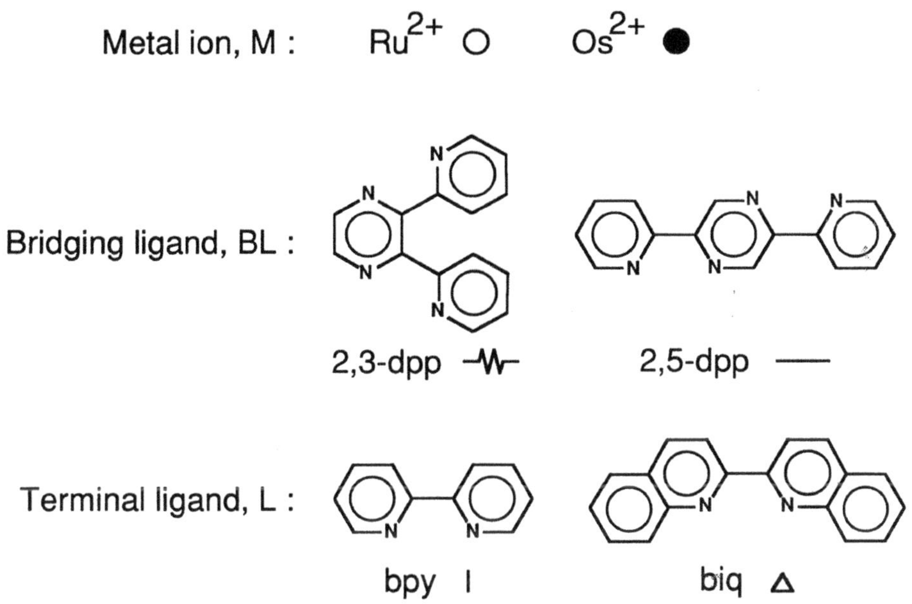

Metal ion, M : Ru^{2+} ○ Os^{2+} ●

Bridging ligand, BL : 2,3-dpp 2,5-dpp

Terminal ligand, L : bpy biq

FIGURE 3-2 Components of the polynuclear complexes and symbols used.

SYNTHETIC STRATEGIES

Oligonuclear Complexes

Mononuclear transition metal complexes are synthesized by combining metal ion (M) and free ligands (L), as shown in eq. (1):

$$M + nL \longrightarrow M(L)_n \qquad (1)$$

In the last few years, we and others have been developing a procedure to synthesize oligonuclear metal complexes of desired nuclearity and chemical structure. Such a procedure is based on the use of *complexes* (building blocks) in the place of the M and/or L in the synthetic reaction (equation 1). The place of M can be taken by mono- or oligonuclear complexes that possess easily replaceable ligands, and the place of L can be taken by mono- or oligonuclear complexes that contain free chelating sites (*complexes as metals* and *complexes as ligands* strategy).

The building blocks used in our syntheses are shown in Table 3-3.[17,24,30-41] Some of them are sketched in Scheme 3-1.

With the exception of the M precursors, the free ligands 2,3-dpp and 2,5-dpp, the complex metals $[Ru(bpy)Cl_3]_n$, $Ru(bpy)_2Cl_2$, and $Ru(biq)_2Cl_2$,

TABLE 3-2 Spectroscopic and Electrochemical Properties of Some $M(L)_n(BL)_{3-n}^{2+}$ Complexes[a]

	Absorption		Luminescence		Electrochemistry			
	λ_{max}[b] (nm)	ϵ (M^{-1} cm^{-1})	λ_{max} (nm)	τ (ns)	$E_{1/2}$(ox) V vs. SCE[c]	$E_{1/2}$(red V vs. SCE[c]		
$Ru(2,3\text{-dpp})_3^{2+}$	455	18,000	623	183	+1.68	−0.95	−1.12	−1.39
$Ru(bpy)_2(2,3\text{-dpp})^{2+}$	475	11,500	691	240	+1.31	−1.06	−1.55	−1.74
$Ru(bpy)_2(2,5\text{-dpp})^{2+}$	486	5,200	695	270	+1.33	−1.03	−1.52	−1.71
$Ru(biq)_2(2,3\text{-dpp})^{2+}$	528	7,600	738	20	+1.47	−0.79	−0.98	−1.40
$Ru(bpy)(2,3\text{-dpp})_2^{2+}$	465	13,600	652[d]	253	+1.46	−0.92	−1.18	−1.58
$Ru(bpy)(2,5\text{-dpp})_2^{2+}$	473	11,200	655[d]	400	+1.39	−0.88	−1.05	−1.55
$Os(2,3\text{-dpp})_3^{2+}$	475	18,000	736[d]	82	+0.92	−0.84	−1.00	−1.36
$Os(2,5\text{-dpp})_3^{2+}$	478	13,000	742[d]	33	+0.96	−0.62	−0.98	−1.32
$Os(bpy)_2(2,3\text{-dpp})^{2+}$	480	10,500	800[d]	20	+0.79	−0.90	−1.39	−1.57
$Os(bpy)_2(2,5\text{-dpp})^{2+}$	498	12,300	824[d]	11	+0.87	−0.90	−1.32	−1.58
$Os(biq)_2(2,3\text{-dpp})^{2+}$	544	14,000	829[d]	—	+1.02	−0.58	−0.79	−1.37
$Os(biq)_2(2,5\text{-dpp})^{2+}$	545	10,600	830[d]	—	+1.04	−0.62	−0.79	−1.23

[a]For the formulas of the ligands, see Figure 3-2; data (refs. 24, 29–32) in acetonitrile solution at room temperature; luminescence lifetimes are in aerated solution.

[b]Lowest-energy ^{1}MLCT band.

[c]Standard saturated calomel electrode.

[d]Uncorrected spectrum.

[e]Not measured because of the low intensity of the emission.

TABLE 3-3 Synthons Employed in the Preparation of Polynuclear Complexes[a]

METALS		LIGANDS	
$RuCl_3.3H_2O$		2,3-dpp	[33]
K_2OsCl_6		2,5-dpp	[34]
		2,3-Medpp$^+$	[35]
COMPLEX METALS		**COMPLEX LIGANDS**	
$[Ru(bpy)Cl_3]_n$	[36]	$[Ru(bpy)_2(2,3\text{-dpp})]^{2+}$	[29, 32]
$Ru(bpy)_2Cl_2$	[37]	$[Ru(biq)_2(2,3\text{-dpp})]^{2+}$	[32]
$Ru(biq)_2Cl_2$	[37]	$[Ru(bpy)_2(2,5\text{-dpp})]^{2+}$	[32]
$Os(bpy)_2Cl_2$	[24]	$[Ru(biq)_2(2,5\text{-dpp})]^{2+}$	[32]
$Os(biq)_2Cl_2$	[24]	$[Os(bpy)_2(2,3\text{-dpp})]^{2+}$	[24]
$Ru(2,3\text{-Medpp})_2Cl_2^{2+}$	[17]	$Ru(2,3\text{-dpp})[(\mu\text{-}2,3\text{-dpp})Ru(bpy)_2]_2^{6+}$	[35, 39]
$Ru[(\mu\text{-}2,3\text{-dpp})Ru(bpy)_2]_2Cl_2^{4+}$	[40]	$[Os(biq)_2(2,3\text{-dpp})]^{2+}$	[24]
$Ru[(\mu\text{-}2,5\text{-dpp})Ru(bpy)_2]_2Cl_2^{4+}$	[40]	$[Os(bpy)_2(2,5\text{-dpp})]^{2+}$	[24]
$Ru[(\mu\text{-}2,3\text{-dpp})Ru(biq)_2]_2Cl_2^{4+}$	[39]	$[Os(biq)_2(2,5\text{-dpp})]^{2+}$	[24]
$Ru[(\mu\text{-}2,3\text{-dpp})Os(bpy)_2]_2Cl_2^{4+}$	[39]	$[Ru(bpy)(2,3\text{-dpp})_2]^{2+}$	[32]
		$[Ru(bpy)(2,5\text{-dpp})_2]^{2+}$	[32]
		$[Ru(2,3\text{-dpp})_3]^{2+}$	[30, 31]
		$[Os(2,3\text{-dpp})_3]^{2+}$	[24]
		$[Os(2,5\text{-dpp})_3]^{2+}$	[24]
		$Ru[(\mu\text{-}2,3\text{-dpp})Ru(bpy)(2,3\text{-dpp})]_3^{8+}$	[41]
		$Ru[(\mu\text{-}2,3\text{-dpp})Ru(2,3\text{-dpp})_2]_3^{8+}$	[17]

[a]Original references in brackets.

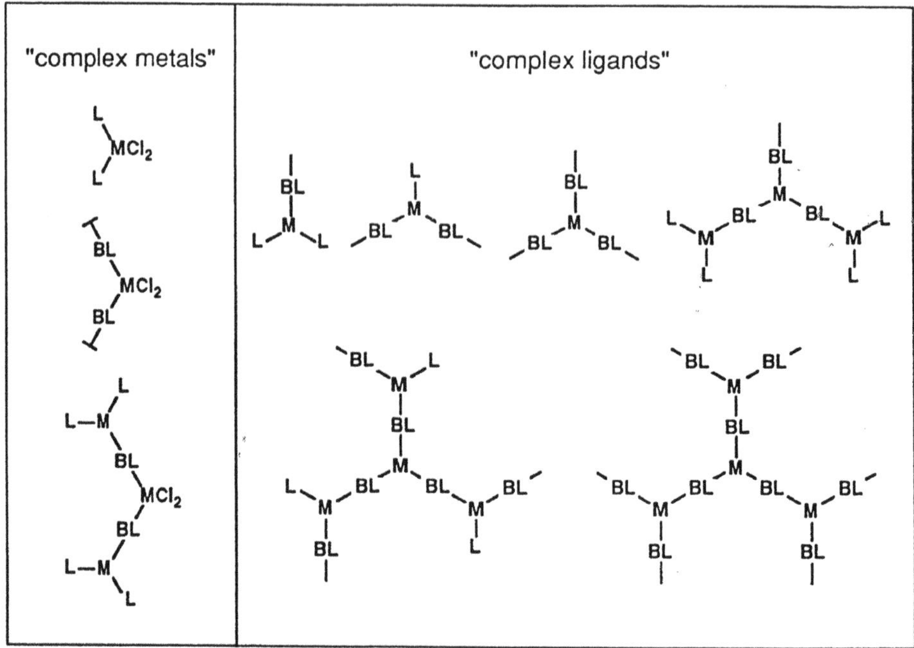

SCHEME 3-1 "Complex metal" and "complex ligand" synthons used to build up complexes of high nuclearity.

and the complex ligands $Ru(bpy)_2(2,3\text{-dpp})^{2+}$ and $Ru(2,3\text{-dpp})_3^{2+}$, all the synthons reported in Table 3-3 were initially prepared in our laboratories. All the charged complexes have been obtained as PF_6^- salts. In the original papers, which are quoted in the table, the synthetic methods can be found, along with characterization data, including elemental analysis, nuclear magnetic resonance (NMR), infrared (IR), visible, and ultraviolet spectra, conductivity values, FAB mass spectra, electrochemical potentials, and luminescence properties. The building blocks are stable in the reaction medium, and no evidence for ligand exchange was obtained.

The synthetic routes followed to obtain complexes of nuclearity from 2 to 22 are illustrated in Table 3-4.[17,24,30,31,38-51] In each equation, the first reactant plays the role of a metal and the second one plays the role of a ligand. A list of the complexes prepared in our laboratory is also reported in the same table. More details on the synthetic procedures (and on the characterization of the compounds) can be found in the original papers.

Schematic views of the of tetranuclear, decanuclear, and tridecanuclear complexes containing 2,3-dpp as a BL are shown in Figures 3-3 to 3-5.

The above-described synthetic strategy is an example of *structure-*

TABLE 3-4 Compounds Prepared and Synthetic Routes Used[a]

DINUCLEAR COMPOUNDS

$$M_a(L_a)_2Cl_2 + M_b(L_b)_2(BL)^{2+} \xrightarrow{1:1} [(L_a)_2M_a(\mu\text{-}BL)M_b(L_b)_2]^{4+}$$

2A	[32, 42]	$M_a = M_b = Ru$	BL = 2,3-dpp	$L_a = L_b$ = bpy
2B	[32]	$M_a = M_b = Ru$	BL = 2,3-dpp	$L_a = L_b$ = biq
2C	[32]	$M_a = M_b = Ru$	BL = 2,3-dpp	L_a = bpy, L_b = biq
2D	[32, 43]	$M_a = M_b = Ru$	BL = 2,5-dpp	$L_a = L_b$ = bpy
2E	[32]	$M_a = M_b = Ru$	BL = 2,5-dpp	$L_a = L_b$ = biq
2F	[32]	$M_a = M_b = Ru$	BL = 2,5-dpp	L_a = bpy, L_b = biq
2G	[24, 38]	$M_a = Ru$, $M_b = Os$	BL = 2,3-dpp	$L_a = L_b$ = bpy
2H	[24]	$M_a = M_b = Os$	BL = 2,3-dpp	$L_a = L_b$ = bpy
2I	[24]	$M_a = M_b = Os$	BL = 2,5-dpp	$L_a = L_b$ = bpy

TRINUCLEAR COMPOUNDS

Method 1

$$M_a(L_a)_2Cl_2 + M_b(L_b)(BL)_2^{2+} \xrightarrow{2:1} [(L_a)_2M_a(\mu\text{-}BL)M_b(L_b)(\mu\text{-}BL)M_a(L_a)_2]^{6+}$$

3A	[44]	$M_a = M_b = Ru$	BL = 2,3-dpp	$L_a = L_b$ = bpy
3B	[44]	$M_a = M_b = Ru$	BL = 2,3-dpp	L_a = biq, L_b = bpy
3C	[44]	$M_a = M_b = Ru$	BL = 2,5-dpp	$L_a = L_b$ = bpy
3D	[44]	$M_a = M_b = Ru$	BL = 2,5-dpp	L_a = biq, L_b = bpy

TABLE 3-4 Continued

Method 2				
$RuCl_3 + M(L)_2(BL)^{2+} \xrightarrow{1:2} Ru[(\mu\text{-}BL)M(L)_2]_2Cl_2^{4+}$				
3E	[40, 45]	M = Ru	BL = 2,3-dpp	L = bpy
3F	[46]	M = Ru	BL = 2,3-dpp	L = biq
3G	[40]	M = Ru	BL = 2,5-dpp	L = bpy
3H	[46]	M = Os	BL = 2,3-dpp	L = bpy
TETRANUCLEAR COMPOUNDS				
Method 1				
$M_a(L)_2Cl_2 + M_b(BL)_3^{2+} \xrightarrow{3:1} M_b[(\mu\text{-}BL)M_a(L)_2]_3^{8+}$				
4A	[31, 47]	$M_a = M_b$ = Ru	BL = 2,3-dpp	L = bpy
4B	[32]	$M_a = M_b$ = Ru	BL = 2,3-dpp	L = biq
4C	[31]	M_a = Ru, M_b = Os	BL = 2,3-dpp	L = bpy
Method 2				
$M_aCl_3 + M_b(L)_2(BL)^{2+} \xrightarrow{1:3} M_a[(\mu\text{-}BL)M_b(L)_2]_3^{8+}$				
4D	[48]	$M_a = M_b$ = Ru	BL = 2,5-dpp	L = bpy
4E	[48]	$M_a = M_b$ = Ru	BL = 2,5-dpp	L = biq
4F	[48]	M_a = Ru, M_b = Os	BL = 2,5-dpp	L = bpy

TABLE 3-4 Continued

Method 3

$M_a[(\mu\text{-}BL_a)M_a(L_a)_2]_2Cl_2^{4+} + M_b(L_b)_2(BL_b)^{2+} \xrightarrow{1:1} [(L_a)_2M_a(\mu\text{-}BL_a)]_2M_a(\mu\text{-}BL_b)M_b(L_b)_2^{8+}$

4GA	[45]	$M_a = M_b = Ru$	$BL = 2,3\text{-dpp}, BL_b = 2,5\text{-dpp}$	$L_a = L_b = bpy$
4H	[49]	$M_a = M_b = Ru$	$BL_a = BL_b = 2,3\text{-dpp}$	$L_a = bpy, L_b = biq$
4I	[49]	$M_a = Ru, M_b = Os$	$BL_a = BL_b = 2,3\text{-dpp}$	$L_a = L_b = bpy$

HEXANUCLEAR COMPOUNDS

Method 1

$M_a[(\mu\text{-}BL_a)M_b(L)_2]_2Cl_2^{4+} + BL_b \xrightarrow{2:1} [(L)_2M_b(\mu\text{-}BL_a)]_2M_a(\mu\text{-}BL_b)M_a[(\mu\text{-}BL_a)M_b(L)_2]_2^{12+}$

6A	[40]	$M_a = M_b = Ru$	$BL_a = BL_b = 2,3\text{-dpp}$	$L = bpy$
6B	[40]	$M_a = M_b = Ru$	$BL_a = BL_b = 2,5\text{-dpp}$	$L = bpy$
6C	[40]	$M_a = M_b = Ru$	$BL_a = 2,3\text{-dpp}, BL_b = 2,5\text{-dpp}$	$L = bpy$
6D	[40]	$M_a = M_b = Ru$	$BL_a = 2,5\text{-dpp}, BL_b = 2,3\text{-dpp}$	$L = bpy$
6E	[39]	$M_a = M_b = Ru$	$BL_a = BL_b = 2,3\text{-dpp}$	$L = biq$
6F	[39]	$M_a = Ru, M_b = Os$	$BL_a = BL_b = 2,3\text{-dpp}$	$L = bpy$

Method 2

$M_a[(\mu\text{-}BL_a)M_b(L_a)_2]_2Cl_2^{4+} + M_a(BL_a)[(\mu\text{-}BL_b)M_a(L_b)_2]_2^{6+}$
$\xrightarrow{1:1} [(L_a)_2M_b(\mu\text{-}BL_a)]_2M_a(\mu\text{-}BL_a)M_a[(\mu\text{-}BL_b)M_a(L_b)_2]_2^{12+}$

6G	[39]	$M_a = M_b = Ru$	$BL_a = 2,3\text{-dpp}, BL_b = 2,5\text{-dpp}$	$L_a = biq, L_b = bpy$
6H	[39]	$M_a = Ru, M_b = Os$	$BL_a = 2,3\text{-dpp}, BL_b = 2,5\ dpp$	$L_a = L_b = bpy$

TABLE 3-4 Continued

HEPTANUCLEAR COMPOUNDS

$$M(L)_2Cl_2 + M[(\mu\text{-}BL)M(L)(BL)]_3^{8+} \xrightarrow{3:1} M[(\mu\text{-}BL)M(L)(\mu\text{-}BL)M(L)_2]_3^{14+}$$

7A	[41]	M = Ru	BL = 2,3-dpp	L = bpy

DECANUCLEAR COMPOUNDS

$$M_a[(\mu\text{-}BL)M_b(L)_2]_2Cl_2^{4+} + M_c(BL)_3^{2+} \xrightarrow{3:1} M_c\{(\mu\text{-}BL)M_a[(\mu\text{-}BL)M_b(L)_2]_2\}_3^{20+}$$

10A	[46, 50]	$M_a = M_b = M_c = Ru$	BL = 2,3-dpp	L = bpy
10B	[46]	$M_a = M_b = M_c = Ru$	BL = 2,3-dpp	L = biq
10C	[46]	$M_a = M_b = Ru$, $M_c = Os$	BL = 2,3-dpp	L = bpy
10D	[46]	$M_a = M_b = Ru$, $M_c = Os$	BL = 2,3-dpp	L = biq
10E	[46]	$M_a = Ru$, $M_b = M_c = Os$	BL = 2,3-dpp	L = bpy
10F	[46]	$M_a = M_c = Ru$, $M_b = Os$	BL = 2,3-dpp	L = bpy

TABLE 3-4 Continued

TRIDECANUCLEAR COMPOUNDS

$$M[(\mu\text{-}BL)M(L)_2]_2Cl_2^{4+} + M[(\mu\text{-}BL)M(L)(BL)]_3^{8+} \xrightarrow{3:1} M\{(\mu\text{-}BL)M(L)(\mu\text{-}BL)M[(\mu\text{-}BL)M(L)_2]_2\}_3^{26+}$$

13A [51] M = Ru BL = 2,3-dpp L = bpy

DOCOSANUCLEAR COMPOUNDS

$$M[(\mu\text{-}BL)M(L)_2]_2Cl_2^{4+} + M[(\mu\text{-}BL)M(BL)_2]_3^{8+} \xrightarrow{6:1} M[(\mu\text{-}BL)M\{(\mu\text{-}BL)M[(\mu\text{-}BL)M(L)_2]_2\}_2]_3^{44+}$$

22A [17] M = Ru BL = 2,3-dpp L = bpy

[a]Original references in brackets.

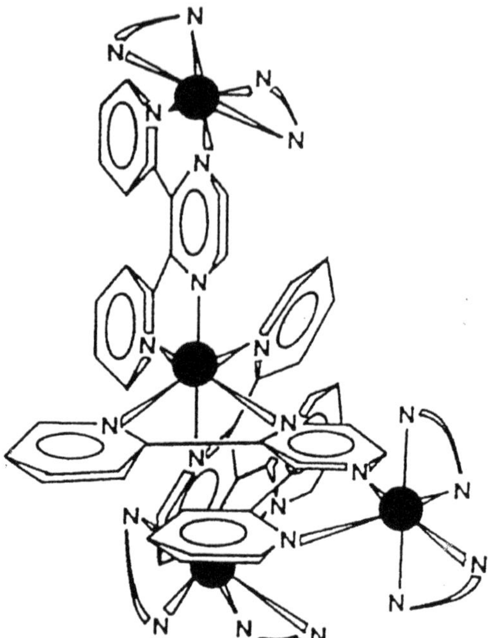

FIGURE 3-3 Schematic view of a tetranuclear complex.

directed synthesis,[52] which allows complicated compounds to be prepared with high yield and satisfactory purity through relatively simple procedures. The crucial step in this synthetic strategy is the design of suitable metal-type and ligand-type building blocks (Table 3-3, Scheme 3-1). Once the building blocks are available, the last step of the synthesis is straightforward and takes place with high yield. For example, for the decanuclear compounds, the last step (Table 3-4) takes place with 70-90% efficiency.[46]

As shown in Table 3-4, the synthetic strategy used is versatile and selective, since the sites occupied by different metals and ligands in the structure of the polynuclear compounds can be predetermined synthetically by an appropriate choice of the building blocks. With the above-described building blocks, however, equivalent supramolecular sites (e.g., the six peripheral ones in the decanuclear complexes, Table 3-4 and Figure 3-4) can be occupied only by the same type of metal ion. With the development of the protection procedure of one of the chelating sites of the BL,[35] it has been possible to synthesize less symmetric and more elongated structures. For instance, two mixed-metal or mixed-ligand hexanuclear complexes have been obtained[39] (see Table 3-4) by employing the trimetallic monofunctional

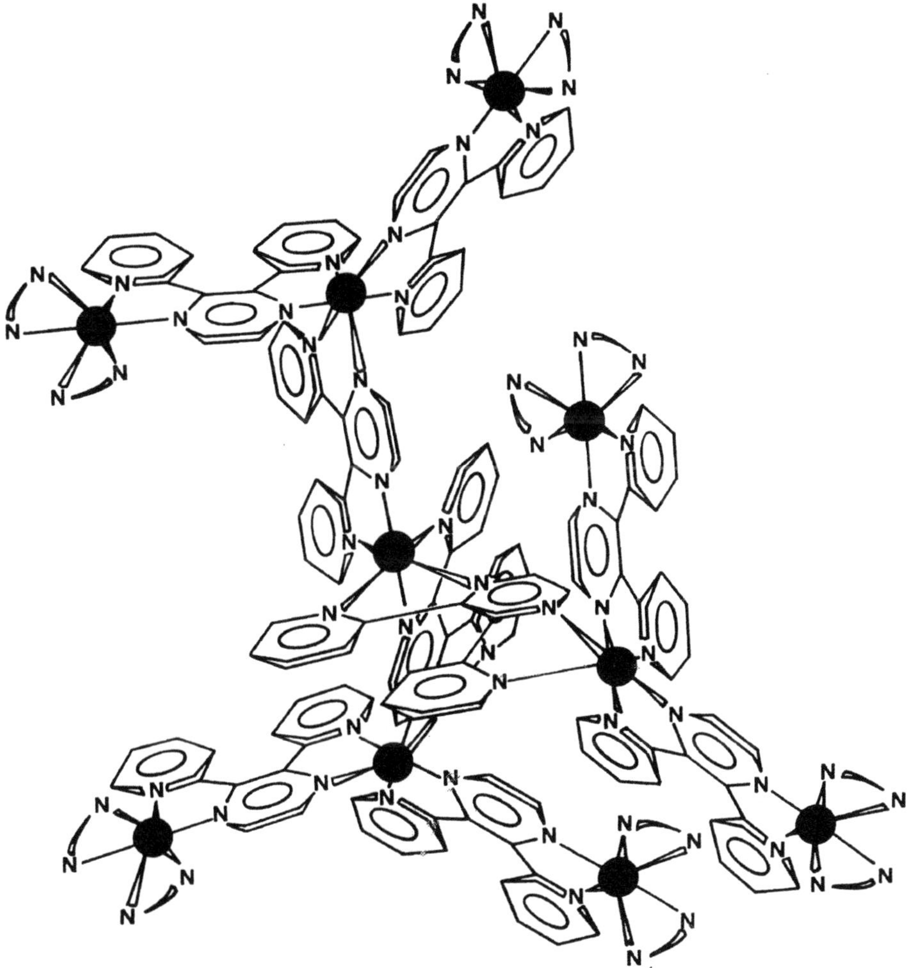

FIGURE 3-4 Schematic view of a decanuclear complex.

synthon $(2,3\text{-dpp})Ru[(\mu\text{-}2,5\text{-dpp})Ru(bpy)_2]_2^{6+}$, whose properties require the use of the protected ligand $2,3\text{-Medpp}^+$ (see below).

Divergent Approach to Arborols

Recently, we have elaborated a strategy to grow oligonuclear metal complexes in ar arborol-like structure using a divergent approach.[17,53]

The synthetic method, illustrated in Scheme 3-2, is based on a protection/deprotection procedure within the previously mentioned complexes-

FIGURE 3-5 Schematic view of a tridecanuclear complex.

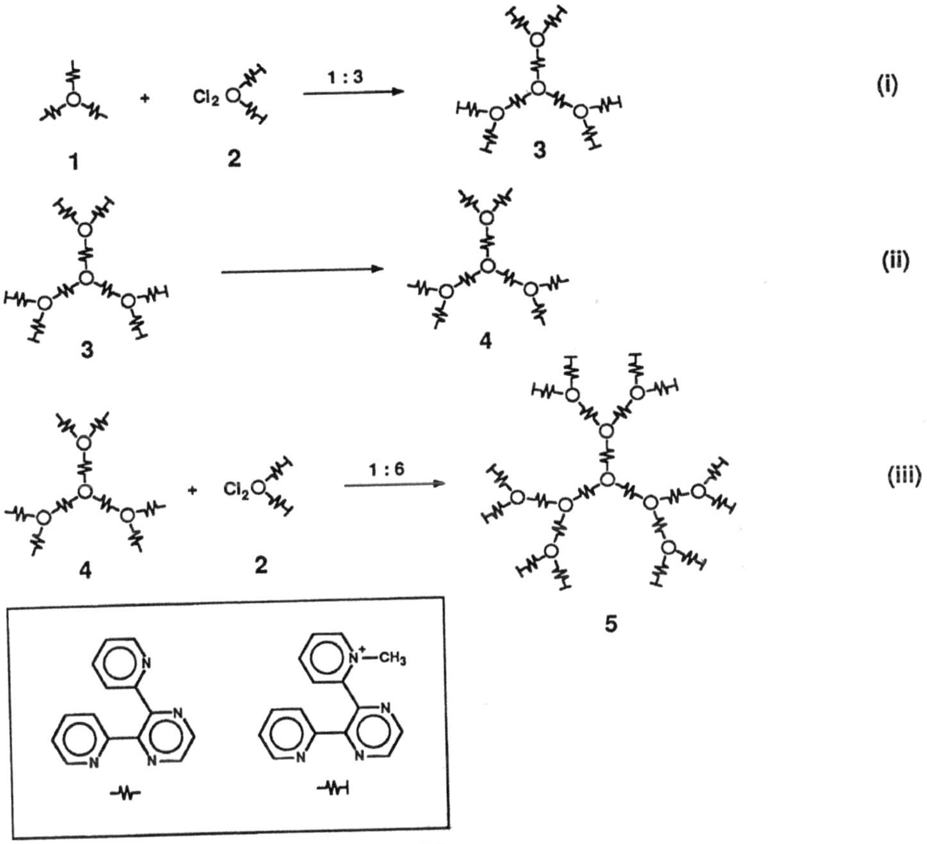

(i)

(ii)

(iii)

SCHEME 3-2

as-ligands/complexes-as-metals synthetic strategy.[54] The first step is the construction of the complex-ligand core $Ru(2,3\text{-dpp})_3^{2+}$ (**1**), which contains three open chelating sites. On each one of these sites, a bifurcate branch can be grafted [Scheme 3-2, eq. (i)] by using the $Ru(2,3\text{-Medpp})_2Cl_2^{2+}$ (**2**) building block, where 2,3-Medpp$^+$ is the 2-[2-(1-methylpyridinium-yl)]-3-(2-pyridyl)pyrazine cation. Compound **2** contains two easily replaced Cl^--ligands (and thus can play the role of a complex metal) and two BL where the chelating sites not used for metal coordination have been protected by methylation. The reaction of **1** with **2** leads to the tetranuclear metal complex **3** ("generation one" of our arborols). Deprotection of the six peripheral chelating sites of **3** [Scheme 3-2, eq. (ii)] yields compound **4**, where six **2** building blocks can again be grafted to yield a "generation two" compound [**5**, Scheme 3-2, eq. (iii)]. In principle, the above procedure can be iterated to obtain larger species by a divergent approach. If the BL of the building block **2** are replaced by simple chelating ligands like bpy, the species obtained from reactions (i) and (iii) are "sterile" since they can no longer be used as cores for larger structures.

The preparation of the docosanuclear compound (Table 3-4),[17] illustrated in Scheme 3-3, shows that the proposed strategy is feasible. Extension to other metal ions [e.g., Os(II)], BL (e.g., 2,5-dpp), and terminal ligands (e.g., 2,2′-biquinoline) is possible.[46]

GENERAL PROPERTIES

The compounds described in this chapter are well-defined supramolecular species made of 2 to 22 metal-containing components. They carry an overall positive charge that is twice the number of the metal atoms. Neglecting the PF_6-counter ions, the docosanuclear compound **22A** (**6** in Scheme 3-3) is made of 1090 atoms, has a molecular weight of 10,890, and has an estimated size of about 5 nm. Besides the 22 metal atoms, it contains 24 terminal bpy ligands and 21 2,3-dpp BL.[17]

All the compounds are soluble in most solvents (e.g., CH_2Cl_2, CH_3CN, C_2H_5OH, H_2O) and are stable both in the dark and upon light excitation.

As we will see later, schematic structures like those shown in Scheme 3-3 are very useful to indicate the chemical composition of the various species and to discuss the interaction between the various building blocks. It should be pointed out, however, that such schemes do not correspond to the real structures of the compounds. Attempts to represent three-dimensional views of some of the complexes examined are shown in Figures 3-3 to 3-5 Structural problems arise because the two coordinating nitrogen atoms of each chelating site of the bridging ligands are not equivalent (Figure 3-2). As a consequence, different geometrical isomers can exist, in principle, for the metal centers coordinated to two or three BL (Figure 3-6). ^{99}Ru and ^{1}H NMR data indicate that for the $[Ru(2,3\text{-dpp})_3]^{2+}$ building block, which is the core of the arborols with 4, 10, 13, and 22 metal-containing units, a 11:1 *mer* vs. *fac* isomer distribution is obtained,[55] but this ratio could change in successive reactions. Each complex can also be a mixture of several diastereoisomeric species owing to the chiral nature of each mononuclear building block. For these reasons, structural investigations on these systems are difficult. Differences arising from the possible presence of isomeric species are not expected to be sizable in the electrochemical and spectroscopic properties discussed below.

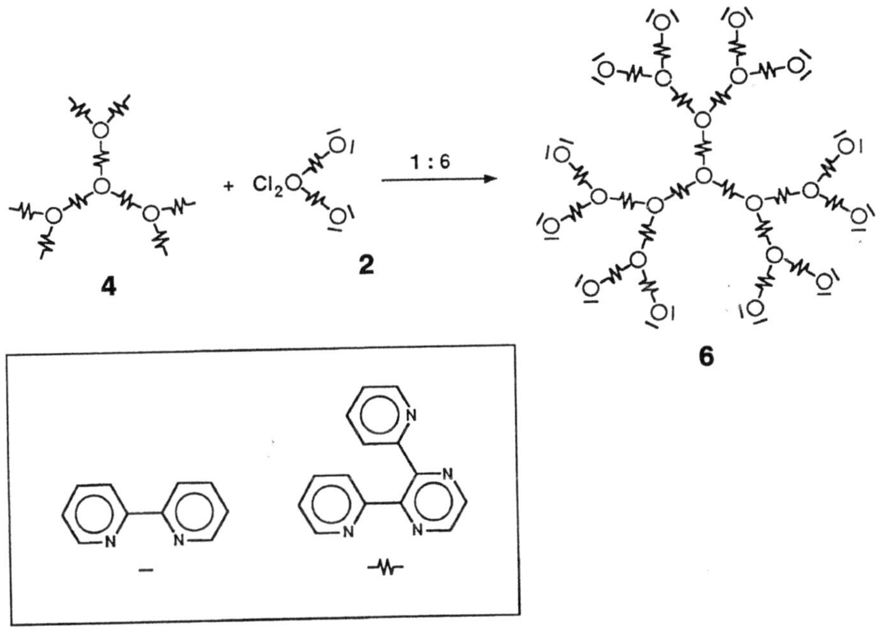

SCHEME 3-3

As one can understand from the schematic views shown in Figures 3-3 to 3-5, the species with high nuclearity exhibit a three-dimensional branching structure of the type shown by otherwise completely different compounds based on polyamidoamines or other organic components.[19,20] Therefore endo- and exoreceptor properties can be expected, which will be the object of future investigations.

We must stress that our complexes differ from most of the organic-type arborols or dendrimers prepared so far for two fundamental reasons: (1) each component exhibits valuable intrinsic properties such as absorption of visible (solar) radiation, luminescence, and redox levels at accessible potentials; (2) by a suitable choice of the building blocks, different components can be placed in specific sites of the supramolecular array, as one can understand from Figure 3-7, where the syntheses of six different decanuclear complexes, using different precursors, are schematized. In other words, our arborols are species with a high "information" content and can therefore be exploited to perform valuable functions (see below).

The absorption, emission, and redox properties of some selected compounds are shown in Table 3-5.[17,31,32,39,41,44,46,51,56]

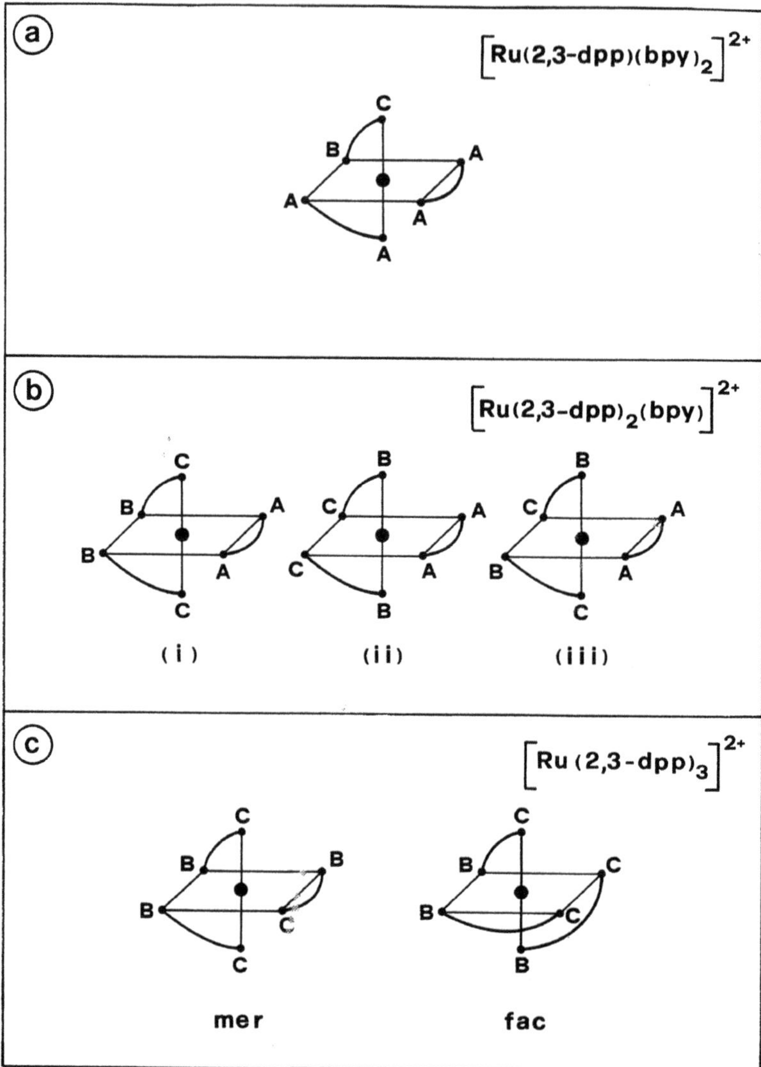

FIGURE 3-6 Possible geometrical isomers for $Ru(2,3\text{-}dpp)_2(bpy)^{2+}$ and $Ru(2,3\text{-}dpp)_3^{2+}$.

ABSORPTION SPECTRA

Since the interaction between the various metal-containing units is weak (see below), the absorption spectra of the oligonuclear complexes are as one can expect from the combination of the absorption spectra of the single metal-containing components. When many components are pres-

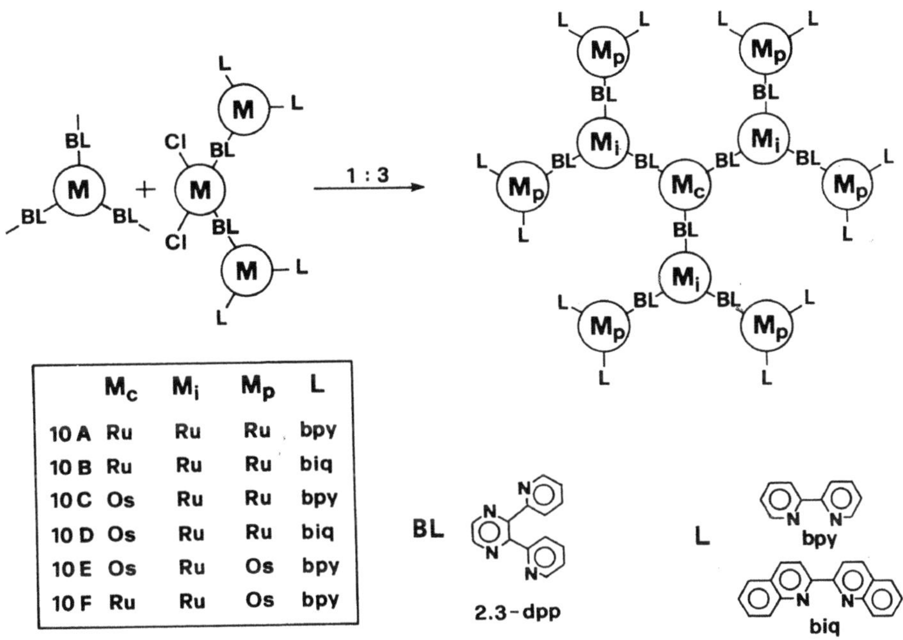

FIGURE 3-7 Scheme showing the synthetic control of the composition of decanuclear compounds.

ent, the absorption spectra show very intense bands throughout the UV and visible spectral region. For example, the spectra of the decanuclear compounds **10B** and **10C** (Figure 3-8)[46] display absorption bands with ϵ up to 600,000 M^{-1} cm^{-1} in the UV region and up to 140,000 M^{-1} cm^{-1} in the visible region. The bands with maxima at 262 and 380 nm can be assigned to $\pi \rightarrow \pi^*$ transitions of the biq ligands, the band at 282 nm to $\pi \rightarrow \pi^*$ transitions on the bpy ligands, and the broad absorption in the 300- to 350-nm region to $\pi \rightarrow \pi^*$ transitions on the biq and 2,3-dpp ligands. The broad bands observed in the visible region receive a contribution from several types of MLCT transitions. The energies of these transitions depend on the nature of the donor metal ion, the acceptor ligand, and the ancillary ligands. Even in the case of the homometallic complex **10B**, five different types of "proximate"[57] MLCT transitions are expected because of the presence of three nonequivalent positions for the metals (M_c, M_i, and M_p) and two nonequivalent positions for the BL (Figure 3-7).

TABEE 3-5 Spectroscopic and Electrochemical Properties of Selected Polynuclear Compounds[a]

		Absorption		Emission		Electrochemistry	
		λ_{max} [b] (nm)	ϵ (M^{-1} cm^{-1})	λ_{max} (nm)	τ (ns)	$E_{1/2}^{ox}$ [n][c] (V vs. SCE)	
2D	$(bpy)_2Ru(\mu\text{-}2,5\text{-}dpp)Ru(bpy)_2^{4+}$	585	11,500	824	155	+1.37 [1]	+1.54 [1]
3A	$(bpy)Ru[(\mu\text{-}2,3\text{-}dpp)Ru(bpy)_2]_2^{6+}$	545	23,500	804	75	+1.48 [2]	
4A	$Ru[(\mu\text{-}2,3\text{-}dpp)Ru(bpy)_2]_3^{8+}$	545	46,000	809	50	+1.50 [3]	
4C	$Os[(\mu\text{-}2,3\text{-}dpp)Ru(bpy)_2]_3^{8+}$	549	40,000	865	18	+1.25 [1]	+1.55 [3]
6D	$[(bpy)_2Ru(\mu\text{-}2,5\text{-}dpp)]_2Ru(\mu\text{-}2,3\text{-}dpp)Ru$ $[(\mu\text{-}2,5\text{-}dpp)Ru(bpy)_2]_2^{12+}$	577	54,100	825	44	+1.36 [4]	
6G	$[(bpy)_2Os(\mu\text{-}2,3\text{-}dpp)]_2Ru(\mu\text{-}2,3\text{-}dpp)Ru$ $[(\mu 2,3\text{-}dpp)Os(bpy)_2]_2^{12+}$	560	81,500	912[d]	[e]	+1.06 [4]	
6H	$[(bpy)_2Os(\mu\text{-}2,3\text{-}dpp)]_2Ru(\mu\text{-}2,3\text{-}dpp)Ru$ $[\mu\text{-}2,3\text{-}dpp)Ru(bpy)_2]_2^{12+}$	576	50,300	760[d] 848[d]	[e] [e]	+1.05 [2]	+1.45 [2]
7A	$Ru[(\mu\text{-}2,3\text{-}dpp)Ru(bpy)(\mu\text{-}2,3\text{-}dpp)Ru(bpy)_2]_3^{14+}$	547	76,200	808	70	+1.38 [3]	
10A	$Ru\{(\mu\text{-}2,3\text{-}dpp)Ru[(\mu\text{-}2,3\text{-}dpp)Ru(bpy)_2]_2\}_3^{20+}$	541	125,000	809	55	+1.43 [6]	

10B	$Ru\{(\mu\text{-}2,3\text{-dpp})Ru[(\mu\text{-}2,3\text{-dpp})Ru(biq)_2]_2\}_3^{20+}$	555	109,000	789	130	+1.62 [6]	
10C	$Os\{(\mu\text{-}2,3\text{-dpp})Ru[(\mu\text{-}2,3\text{-dpp})Ru(bpy)_2]_2\}_3^{20+}$	550	117,000	808 860	65	+1.17 [1]	+1.50 [6]
10F	$Os\{(\mu\text{-}2,3\text{-dpp})Ru[(\mu\text{-}2,3\text{-dpp})Os(bpy)_2]_2\}_3^{20+}$	563	140,500	900[d]		+1.05 [6]	+1.39 [1]
13A	$Ru\{(\mu\text{-}2,3\text{-dpp})Ru(bpy)(\mu\text{-}2,3\text{-dpp})Ru[(\mu\text{-}2,3\text{-dpp})Ru(bpy)_2]_2\}_3^{26+}$	544	133,000	800	62	+1.50 [9][f]	
22A	$Ru\{(\mu\text{-}2,3\text{-dpp})Ru[(\mu\text{-}2,3\text{-dpp})Ru\{(\mu\text{-}2,3\text{-dpp})Ru(bpy)_2\}_2]_2\}_3^{44+}$	542	202,400	802		+1.52 [12]	

[a]Experiments in acetonitrile solution at room temperature unless otherwise noted. Luminescence lifetimes are aerated values unless otherwise stated. Data are taken from refs. 32 (2A), 44 (3A), 31 (4A and 4C), 39 (6D and some data on 6G and 6H), 56 (luminescence data on 6G, 6H, and 10F), 41 (7A), 46 (10A, 10B, 10C, and 10F), 51 (13A), and 17 (22A).

[b]Lowest-energy ^{1}MLCT band.

[c]The numbers in brackets indicate the number of electrons exchanged. For details, see the original reference.

[d]At 90 K in MeOH/EtOH 4:1 (v/v) glassy matrix.

[e]Not measured.

[f]This wave exhibits a large separation between cathodic and anodic peaks (ΔE = 180 mV). This suggests that there is a superposition of two closely spaced oxidation processes; the first has been attributed to one-electron oxidation of six independent redox sites and the second to one-electron oxidation of three independent redox sites. For more details, see the original reference.

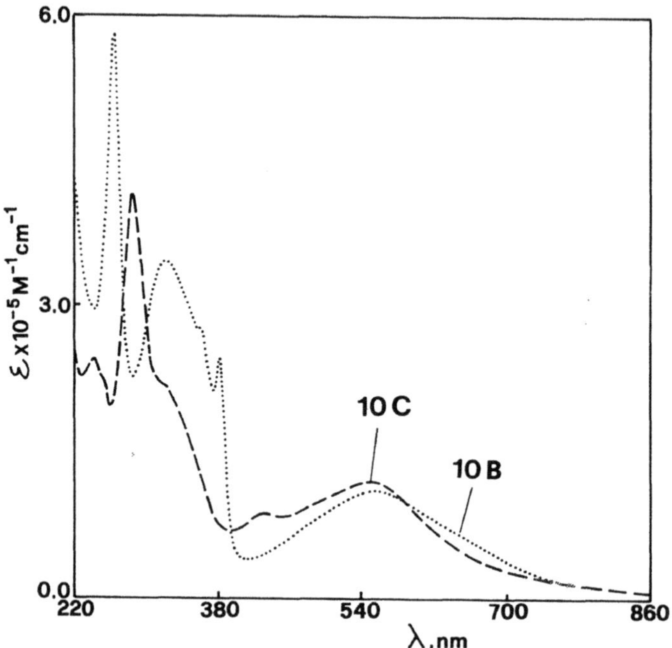

FIGURE 3-8 Absorption spectra of the decanuclear compounds **10B** and **10C** in acetonitrile solution at room temperature.

From the spectroscopic and electrochemical (see below) data, the following trends are observed for the energy ordering of the metal-to-ligand charge transfer transitions: (1) for the same acceptor ligand and the same ancillary ligands, Os^{2+} → acceptor ligand < Ru^{2+} → acceptor ligand; (2) for the same metal and ancillary ligands, M → BL < M → biq < M → bpy; (3) for the same metal and the same acceptor ligands, $(bpy)_2M$ → acceptor ligand < $(biq)_2M$ → acceptor ligand < $(BL)_2M$ → acceptor ligand. From a comparison of the absorption spectra of **10B** and **10C** (Figure 3-8), it is easy to assign the absorption band at about 430 nm to M → bpy CT transitions. The broad band with maximum at ~550 nm receives contributions from M → biq and M → BL CT transitions. It should be noted that in the Os-containing compounds (particularly in **10E** and **10F**), the absorption bands extend to the red because the probability of the singlet → triplet transitions is relatively high owing to the enhanced spin-orbit coupling. The above discussion of the absorption spectra of the decanuclear complexes can be extended directly to the other complexes of this family.

ELECTROCHEMICAL PROPERTIES

As mentioned above, the mononuclear compounds of the $M(BL)_{3-n}(L)_n^{2+}$ series exhibit reversible redox processes. On reduction, each L ligand is reduced twice and each BL ligand is reduced four times in the potential window $-0.5/-3.1$ V.[58] The reduction potential of each ligand depends on its electronic properties and, to a smaller extent, on the nature of the metal and of the other ligands coordinated with the metal. It follows that on reduction of a polynuclear complex that contains several ligands (e.g., 21 ligands are present in the decanuclear complexes), an extremely complicated pattern with many overlapping peaks is observed. On oxidation of mononuclear complexes in the potential window $< +1.6$ V, only one peak is observed in the differential pulse voltammograms (Table 3-2), which concerns the metal ion. The oxidation potential depends strongly on the nature of the metal ion [Os(II) is oxidized at less positive potentials compared to Ru(II)] and, less dramatically, on the nature of the coordinated ligands. Because of the previously discussed electronic properties of the isolated components (and of the stabilization of the LUMO of 2,3-dpp and 2,5-dpp on coordination with a second metal center), it can be expected that for the metal-containing building blocks present in the synthesized polynuclear compounds (Table 3-4), the oxidation potential of the metal [and the energy of the lowest (luminescent) excited state] increases in the series $Os(bpy)_2(\mu\text{-}2,5\text{-dpp})^{2+} \leq Os(bpy)_2(\mu\text{-}2,3\text{-dpp})^{2+} < Os(biq)_2(\mu\text{-}2,5\text{-dpp})^{2+} \leq Os(biq)_2(\mu\text{-}2,3\text{-dpp})^{2+} < Os(\mu\text{-}2,5\text{-dpp})_3^{2+} < Os(\mu\text{-}2,3\text{-dpp})_3^{2+} < Ru(bpy)_2(\mu\text{-}2,5\text{-dpp})^{2+} \leq Ru(bpy)_2(\mu\text{-}2,3\text{-dpp})^{2+} < Ru(biq)_2(\mu\text{-}2,5\text{-dpp})^{2+} \leq Ru(biq)_2(\mu\text{-}2,3\text{-dpp})^{2+} < Ru(bpy)(\mu\text{-}2,5\text{-dpp})_2^{2+} \leq Ru(bpy)(\mu\text{-}2,3\text{-dpp})_2^{2+} < Ru(\mu\text{-}2,3\text{-dpp})_3^{2+}$.

It can also be expected (and it is confirmed by the experimental data; see Table 3-5) that in the polynuclear complexes the metal-metal interaction is nonnegligible for metals coordinated with the same BL, whereas it is very weak for metals that do not share the same BL.

It is well known[59,60] that supramolecular species containing a number of identical noninteracting centers exhibit current-potential responses having the same shape as that obtained with the corresponding molecule containing a single center. Only the magnitude of the current is enhanced by the presence of additional electroactive centers. Thus, equivalent units that are not directly connected are oxidized at the same potential. This allows us to control the number of electrons lost at certain potential by placing in the supramolecular species the desired number of suitable, equivalent, and noninteracting units. For example, if we wish to have a species that undergoes a three-electron oxidation process at the same potential, we can take a tetranuclear complex like **4A**[31] (Table 3-5): the three peripheral, equivalent, and noninteracting $Ru(bpy)_2(\mu\text{-}2,3\text{-dpp})^{2+}$ units are indeed oxidized at the same potential, +1.50 V. This three-electron peak should be followed (outside the potential window examined) by a one-electron peak related to the oxidation of the central $Ru(\mu\text{-}2,3\text{-dpp})_3^{2+}$ component (such a peak is expected to be strongly displaced to more positive potentials because of the presence of the three contiguous, already oxidized peripheral components). If we wish a 1-3 oxidation pattern (i.e., a one-electron peak followed by a three-electron peak), we can select complex **4C.**[31] In this complex, the central, Os-containing unit is oxidized at +1.25 V, followed by simultaneous oxidation of the three peripheral, equivalent, and noninteracting $Ru(bpy)_2(\mu\text{-}2,3\text{-dpp})^{2+}$ components at +1.55 V (Table 3-5). A different oxidation pattern (namely, a 1-2 one) is obtained by using the tetranuclear complex **4I:**[60] the peripheral Os-containing unit is oxidized at +1.00 V, followed by simultaneous oxidation of the two peripheral Ru-containing units at +1.50 V.

The "symmetrical" hexanuclear compounds **6D** and **6G** exhibit a four-electron oxidation peak that can be attributed to oxidation, at nearly the same potential, of the four equivalent peripheral metal ions (Table 3-5, Figure 3-9).[40] The oxidation peaks of the two inner metal ions are expected to occur at higher potential; they are further displaced to more positive values owing to the presence of the already oxidized peripheral units, and thus these peaks cannot be observed in the potential window examined. For compound **6H,** however, the oxidation pattern is quite different: two peaks of equal height, both corresponding to a two-electron oxidation process, are observed (Figure 3-9).[39] The first oxidation peak occurs at nearly the same potential as the four-electron peak of compound **6G.** This shows that, as expected, the two Os-containing units are the first to be oxidized. The second peak concerns the oxidation of the two $Ru(bpy)_2(\mu\text{-}2,5\text{-dpp})^{2+}$ units. Since such units lie far away from the previously oxidized Os-containing units, their oxidation potential is close to that of the equivalent peripheral units of **6D.**

Very interesting oxidation patterns are obtained for the decanuclear compounds (Table 3-5, Figure 3-7).[46] In compound **10C,** which contains one Os^{2+} and nine Ru^{2+} ions, the Os^{2+} ion is expected to be oxidized at less positive potentials than the nine Ru^{2+} ions. Furthermore, because of the different electron donor properties of the ligands, the six peripheral Ru^{2+} ions are expected to be oxidized at less positive potentials than the three intermediate Ru^{2+} ions. The differential pulse voltammogram of **10C** (Figure 3-10) shows an oxidation peak at +1.17 V that is in the expected range for Os^{2+} oxidation, and can therefore be assigned to one-electron oxidation of the central Os^{2+} metal ion, and another peak at +1.50 V that has the same bandwidth but a six times higher intensity.

This peak can be assigned to the independent one-electron oxidation of the six peripheral noninteracting Ru^{2+} ions. Oxidation of the three intermediate Ru^{2+} ions is further shifted toward more positive potentials and cannot be observed in the potential window examined.

For **10E**, one expects that oxidation involves first the six peripheral Os^{2+} ions (which contain the stronger electron donor bpy ligand in their coordination sphere) and then the central one. This is fully consistent with the differential pulse voltammetry results (Figure 3-10). The more positive oxidation potential of the central Os^{2+} ion in **10E** compared to those of the same metal ions in **10C** is obviously due to the presence of six already oxidized osmium ions when the central ion of **10E** undergoes oxidation.

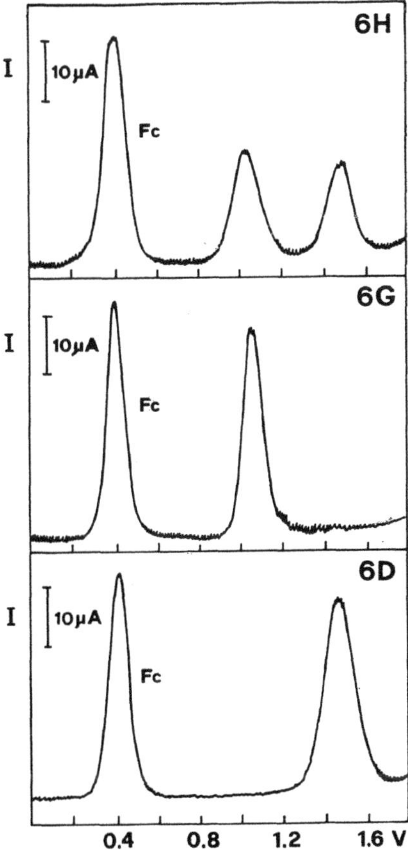

FIGURE 3-9 Oxidation pattern for some hexanuclear complexes. Fc indicates the oxidation peak of ferrocene, used as an internal standard.

For **10A** only one oxidation peak is observed, so that internal comparison cannot be made. On the other hand, evaluation of the number of electrons involved in this peaks by *direct* comparison with the area of the ferrocene oxidation peak would require a correction for the very different diffusion coefficients of ferrocene and of the decanuclear species. Such a procedure can be avoided by keeping the same ferrocene concentration in each experiment, so that it is possible to normalize the areas of the peak of complex **10A** to those of complexes **10C** and **10E** by using the peak of ferrocene as a comparison. The value so obtained shows that the oxidation peak of **10A** corresponds to the one-electron oxidation of six noninteracting units and can thus be assigned to the oxidation of the

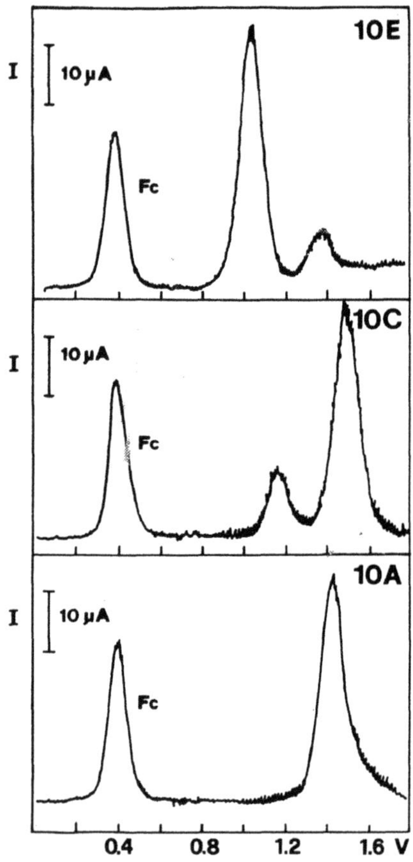

FIGURE 3-10 Oxidation patterns for some decanuclear complexes. Fc indicates the oxidation peak of ferrocene used as an internal standard.

peripheral metal ions. Oxidation of the central and intermediate metal ions cannot be observed in the potential window examined.

As far as the tridecanuclear compound **13A** (Figure 3-5) is concerned,[51] the expected oxidation pattern consists of a six-electron peak regarding the independent one-electron oxidation of the six peripheral Ru-containing units, followed at slightly more positive potential by a three-electron peak due to one-electron oxidation of the three $Ru(\mu\text{-}2,3\text{-}dpp)_2(bpy)^{2+}$ units. Oxidation of the remaining four metal ions is expected to occur out of the examined potential window. The experimental results are as follows: a broad, reversible oxidation peak is observed at +1.50 V, with an area corresponding to nine electrons and a half-width of 170 mV (a value much larger than that expected for oxidation of noninteracting units). Clearly, the oxidation processes involving the six $Ru(\mu\text{-}2,3\text{-}dpp)(bpy)_2^{2+}$ units and the three $Ru(\mu\text{-}2,3\text{-}dpp)_3^{2+}$ units occur at potentials too close to be resolved.

In conclusion, the electrochemical data offer a fingerprint of the chemical and topological structure of the oligonuclear compounds. Furthermore, made-to-order synthetic control of the number of electrons exchanged at a certain potential can be achieved. Examination over a more extended oxidation potential window (in a solvent like liquid SO_2) will allow us to obtain a larger variety of oxidation patterns.

LUMINESCENCE

The polynuclear complexes discussed in this chapter exhibit intense absorption bands in the UV and visible spectral regions, as is shown by the spectra of the decanuclear complexes **10B** and **10C** displayed in Figure 3-8. As we have seen, in these compounds metal-metal interaction is weak for metals coordinated with the same BL and almost negligible for metals that are far apart. Similarly, ligand-ligand interaction is appreciable only for ligands coordinated with the same metal. As a consequence, the energy levels of the component building blocks are essentially maintained. Light excitation in the visible absorption bands populates ^{1}MLCT excited states in the various components. Investigations carried out on $Ru(bpy)_3^{2+}$ with fast techniques indicate that the originally populated ^{1}MLCT excited states undergo relaxation to the lowest-energy ^{3}MLCT level in the subpicosecond time scale.[61] If this behavior, as it seems likely, is of general validity for the various components of the polynuclear compounds, the actual result of light excitation is the population with unitary efficiency of the lowest-energy ^{3}MLCT level of the component where light absorption has taken place. If each component were isolated, as happens in mononuclear complexes, competition between radiative (luminescence) and radiationless decay to the ground state would account for the deactivation of the ^{3}MLCT level, with an overall rate constant, measured from the luminescence decay, in the range 10^6 to 10^8 s^{-1}. All the members of the mononuclear $M(BL)_{3-n}(L)_n^{2+}$ family, in fact, display a characteristic luminescence both in rigid matrix at 77 K and in fluid solution at room temperature (Table 3-2). When the components are linked together in a supramolecular array, electronic energy can be transferred from an excited component to an unexcited one even if the electronic interaction is weak. In most of the examined polynuclear compounds only a luminescence band, corresponding to the lowest-energy ^{3}MLCT level, is observed (Table 3-5), indicating that energy transfer from upper-lying to lower-lying levels does occur (see below).

ANTENNA EFFECT

The natural photosynthetic systems[62] show that for solar energy conversion purposes, supramolecular arrays are needed that absorb as much visible light as possible and are capable of channeling the resulting excitation energy toward a specific site of the array (*antenna devices*). The polynuclear metal complexes described in this chapter are excellent light absorbers in the entire visible region. Furthermore, efficient energy transfer can take place between their components, as shown by the presence of only one luminescence band for compounds that contain more than one type of chromophoric unit Quantitatively, the occurrence of energy transfer can be established by the quenching of the luminescence of the donor unit and the sensitization of the luminescence of the acceptor unit, and its efficiency can be estimated by comparing the absorption and excitation spectra. The rate of energy transfer depends on the energetics of the process and on the electronic interaction between the components.[63] In polynuclear Ru(II) and Os(II) polypyridine complexes,[64-66] both the coulombic and exchange energy transfer mechanisms may play a role, but the exchange mechanism (involving the lowest excited state of the donor) seems to be more effective, especially for short separation distances.[63] Exchange energy transfer between two different ($M_1 \rightarrow L_1$ and $M_2 \rightarrow L_2$) CT excited states of a polynuclear complex can be viewed as a two-electron transfer process, $M_2 \rightarrow M_1$ and $L_1 \rightarrow L_2$.

The electrochemical experiments discussed above show that the metal-metal interaction through a 2,3- or 2,5-dpp bridge is not negligible.[32] Therefore, energy transfer between *directly connected* units can take place by an adiabatic mechanism. Since the reorganizational energy is small, intercomponent energy transfer between such units is expected to be very fast whenever it is exoergonic. As we will see later, the experimental results indicate that the efficiency of energy transfer for exoergonic processes between components connected by a BL is usually 100% in these systems. Exoergonic energy transfer between remote components, however, may well be slow because of a poor electronic interaction.

On the basis of the above considerations, it has been possible to design polynuclear complexes in which the electronic energy generated by light absorption can be channeled toward a desired direction along the supramolecular structure.[46,49,67] This can be done by locating different building blocks in an appropriate sequence so as to have a driving force for energy transfer.

As we have seen, the energy levels involved in energy transfer are the lowest ^{3}MLCT excited state of each component. Since the pioneering work of Crosby,[68] it is known that as Os(II) is easier to oxidize than Ru(II), the luminescent level of any Os(II)-polypyridine complex lies lower in energy than the luminescent level of the corresponding Ru(II) complex. We have also seen that for complexes of the same metal, the energy of the luminescent level can be tuned by changing the type of polypyridine ligand (Table 3-1).[21,68] A third, indirect way to tune the energy of the luminescent levels is to change the nature of the ancillary ligands (i.e., of the ligands that are not involved in the lowest-energy MLCT transition). For example, when an X ligand of an ML_nX_m complex is replaced by a Y ligand that is a better electron donor that X, the ^{3}MLCT level will move to lower energy because of the increased electron density (and the consequent decrease in the ionization potential) of the metal. As previously mentioned, extensive work on Ru^{2+} and Os^{2+} complexes has shown that the 2,3-dpp and 2,5-dpp BL and the bpy and biq terminal ligands can be ordered in the following series as far as the electron-donor power is concerned: bpy > biq > μ-2,3-dpp ≥ μ-2,5-dpp. This means, for example, that coordination of bpy with Ru^{2+} reduces the positive charge of the metal ion more than coordination of μ-2,3-dpp. On the other hand, the energy of the lowest (empty) π^* orbital increases in the series μ-2,5-dpp < μ-2,3-dpp < biq < bpy. This means, for example, that in a $Ru(bpy)_2(\mu\text{-}2,3\text{-dpp})^{2+}$ moiety the Ru → (μ-2,3-dpp) excited state lies lower in energy than the Ru → bpy excited state.

Because of these properties of the components and of the stabilization of the LUMO of 2,3-dpp and 2,5-dpp on coordination with a second metal center, it can be expected that for the metal-containing building blocks present in the polynuclear compounds, the energy of the lowest (luminescent) excited state increases (as does the oxidation potential; see above) in the series $Os(bpy)_2(\mu\text{-}2,5\text{-dpp})^{2+}$ ≤ $Os(bpy)_2(\mu\text{-}2,3\text{-dpp})^{2+}$ <

$Os(biq)_2(\mu\text{-}2,5\text{-}dpp)^{2+}$ ≤ $Os(biq)_2(\mu\text{-}2,3\text{-}dpp)^{2+}$ < $Os(\mu\text{-}2,5\text{-}dpp)_3^{2+}$ < $Os(\mu\text{-}2,3\text{-}dpp)_3^{2+}$ < $Ru(bpy)_2(\mu\text{-}2,5\text{-}dpp)^{2+}$ ≤ $Ru(bpy)_2(\mu\text{-}2,3\text{-}dpp)^{2+}$ < $Ru(biq)_2(\mu\text{-}2,5\text{-}dpp)^{2+}$ ≤ $Ru(biq)_2(\mu\text{-}2,3\text{-}dpp)^{2+}$ < $Ru(bpy)(\mu\text{-}2,5\text{-}dpp)_2^{2+}$ ≤ $Ru(bpy)(\mu\text{-}2,3\text{-}dpp)_2^{2+}$ < $Ru(\mu\text{-}2,3\text{-}dpp)_3^{2+}$.

By using the above guidelines, it is possible to design polynuclear complexes in which the component(s) with the lowest energy excited state(s) is (are) located in the desired position(s) of the supramolecular structure. As discussed above, this allows synthetic control of the direction(s) of energy migration after light absorption. A few examples will now be presented.

The four possible energy migration patterns for tetranuclear compounds are schematized in Figure 3-11.[49] Pattern 4-I obtains for L = bpy, BL = 2,3-dpp, and M = Ru^{2+} (e.g., for compound **4A**, Table 3-4). In such a complex, the three peripheral units are equivalent and their lowest excited state lies at lower energy than the lowest excited state of the central unit. The luminescence of the central unit is quenched, and that of the peripheral ones is sensitized. The efficiency of energy transfer is difficult to estimate because of strong band overlap in the absorption (and excitation) spectrum. Pattern 4-II is found for L = bpy, BL = 2,3-dpp, $M_4 = Os^{2+}$, $M_1 = M_2 = M_3 = Ru^{2+}$ (compound **4C**). The luminescence of the peripheral units is quenched, and the luminescence of the central unit is sensitized with ~100% efficiency.

To channel energy toward a single peripheral unit (pattern 4-III), we have designed and synthesized a compound with L = bpy, BL_1 = 2,5-dpp, $BL_2 = BL_3$ = 2,3-dpp, M = Ru^{2+} (compound **4G**). Since 2,5-dpp is easier to reduce than 2,3-dpp, the Ru → BL CT excited state of the M_1-based building block is lower in energy than the Ru → BL CT state of the building blocks based on M_2 and M_3. On the other hand, the lowest excited state of the central building block is higher in energy than the lowest excited state of all the peripheral ones. Thus, energy transfer from the M_4-, M_2-, and M_3-based building blocks to the M_1-based one is exoergonic, but the energy transfer process from the M_2- and M_3-based units to the M_1-based unit must either overcome a barrier at M_4 or proceed through space. The luminescence results show that the only emitting level is that based on M_1 and that energy migration is ~100% efficient.

To obtain the migration pattern illustrated by 4-IV, we have designed and synthesized a compound with L_1 = biq, $L_2 = L_3$ = bpy, BL = 2,3-dpp, and M = Ru^{2+} (**4H**). Since the energy level of the M_4-based building block is (slightly) higher than that of the M_1-based one, the migration process must again overcome an energy barrier. The results obtained with this complex show that energy migration takes place with high efficiency.

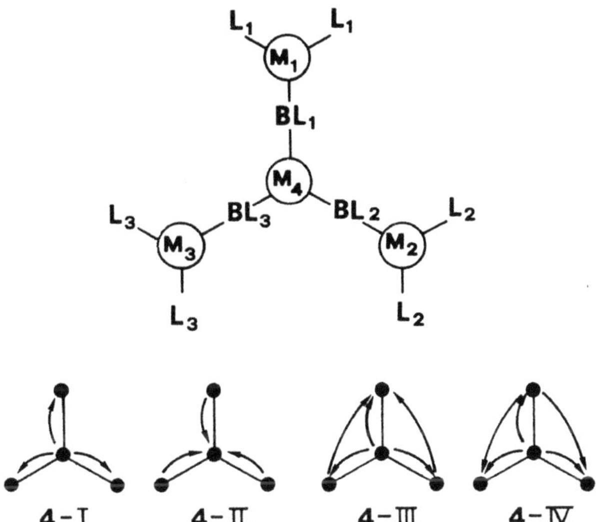

FIGURE 3-11 Energy migration patterns for tetranuclear compounds; for more details, see text.

For the decanuclear compounds **10A** to **10F** (Table 3-4), the directions along which energy transfer processes are exoergonic are schematically indicated by arrows in Figure 3-12.[46] Compound **10A** displays a luminescence band at 809 nm that can be straightforwardly assigned to the peripheral $(bpy)_2Ru \rightarrow$ BL excited states. The lack of luminescence at shorter wavelengths and the constancy of the luminescence quantum yield on changing the excitation wavelength indicate that the chromophoric groups based on the central and intermediate Ru^{2+} ion undergo efficient deactivation to the peripheral Ru-based units, as expected, because energy transfer is exoergonic in the direction from center to periphery. Compound **10B** exhibits exactly the same behavior as **10A.** For compounds **10C** and **10D,** a broad luminescence band is observed at room temperature, with a shoulder on its low energy tail. The maxima of the luminescence bands almost coincide with those of the bands exhibited by **10A** and **10B,** respectively. The predominant emission can thus be assigned to the peripheral units. Subtraction (after normalization) of the luminescence band of **10A** from that of **10C** and of the luminescence band of **10B** from that of **10D** yields a band with a maximum at ~860 nm, as expected for the luminescence of a central Os unit. We conclude that, at room temperature, **10C** and **10D** emit from both the central and peripheral units. Such behavior is consistent with the fact that in **10C** and **10D** the lowest excited state of the intermediate Ru-based units lies at higher energy (~2000 cm^{-1}) than the lowest excited state of the peripheral units. Thus, for **10C** and **10D**, the two-step energy transfer process from the peripheral units to the central one (where the lowest-energy excited state of the supramolecular array is located) must be very slow, since its first step is endoergonic by ~2000 cm^{-1}. Direct (through space) energy transfer from the peripheral to the central units is exoergonic but should be slow because of the large separation distance. A quantitative evaluation of the energy transfer efficiency from the peripheral to the central unit is difficult to make from luminescence quantum yield data because of the strong overlap between the absorption bands of the various units.

For **10E** and **10F** the lowest excited states are localized on the peripheral $(bpy)_2Os \rightarrow BL$ units, which are expected to emit around 900 nm. With IR-sensitive equipment,[56] luminescence bands at 900 and 892 nm can in fact be observed for **10E** and **10F**, respectively. The lack of any Ru-based luminescence for **10F** indicates a 100% efficient center-to-periphery channeling of the excitation energy, as expected because of the energy gradient (Figure 3-12). For **10E**, deactivation of the central Os-based unit by the peripheral ones should not occur because the first step of this process is endoergonic (Figure 3-12). The lack of observable luminescence from such a central unit may simply be due to the fact that most of the

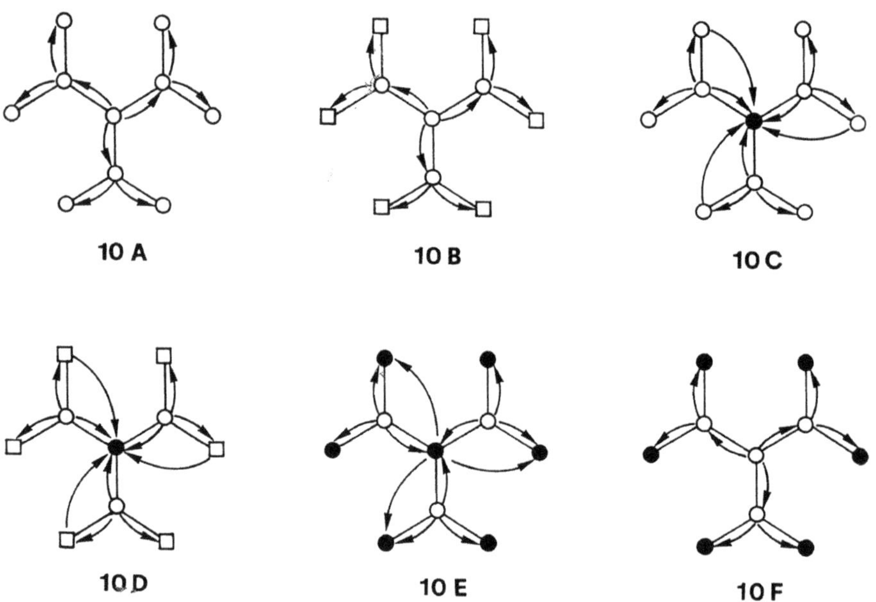

FIGURE 3-12 Energy migration patterns for some decanuclear compounds. Empty and full circles indicate Ru^{2+} and Os^{2+}, respectively. In the peripheral positions, circles and squares indicate $M(bpy)_2$ and $M(biq)_2$ components, respectively. The arrows indicate the exoergonic energy transfer steps.

light at any excitation wavelength is absorbed by the much more numerous peripheral and intermediate units.

CONCLUSION

A synthetic strategy has been developed to obtain supramolecular dendritic structures of nanometric dimensions made of metal complex units capable of high information content (light absorption, luminescence, redox activity). Specific metal and/or ligands can be placed in predetermined sites of the supramolecular array by an appropriate choice of the building blocks. It is thus possible to design species in which several important functions can be synthetically controlled. In particular, made-to-order control of the number of electrons lost at a certain potential and of the direction of electronic energy transfer can be achieved.

Because of their strong absorption in the visible spectral region and the possibility of predetermining the direction of energy migration, these compounds can be used as photochemical molecular devices (e.g., as antennas for harvesting solar energy).[69,70] Because of the presence of several interacting and/or noninteracting redox centers, they are good candidates for the role of multielectron-transfer catalysts.

Acknowledgments

We wish to thank V. Cacciari, G. Gubellini, L. Minghetti, and F. Stillitani for technical assistance. This work was supported by the Consiglio Nazionale delle Ricerche (Progetto Finalizzato Chimica Fine II) and Ministero dell'Università e della Ricerca Scientifica e Tecnologica.

REFERENCES

1. a. *Physics and Technology of Submicron Structures,* H. Heinrich, G. Bauer, and F. Kuchar, Eds. (Springer, Berlin, 1988). b. H.I. Smith and H.G. Craighead, *Phys. Today,* **43** (2), 24 (1990).
2. a. *Molecular Electronic Devices,* F.L. Carter, R.E. Siatkowski, and H. Woltjen, Eds. (North Holland, Amsterdam, 1988). b. K.E. Drexler, *Engines of Creations* (Fourth Estate, London, 1990). c. P. Ball and L. Garwin, *Nature,* **355,** 761 (1992).
3. J.-M. Lehn, *Angew. Chem. Int. Ed. Engl.,* **29,** 1304 (1990).
4. J.-P. Sauvage, *Acc. Chem. Res.,* **23,** 319 (1990).
5. F. Vögtle *Supramolecular Chemistry* (Wiley, Chichester, 1991).
6. *Frontiers in Supramolecular Organic Chemistry and Photochemistry,* H.-J. Schneider and H. Dür, Eds. (VCH, Weinheim, 1991).
7. V. Balzani and F. Scandola, *Supramolecular Photochemistry* (Horwood, Chichester, 1991).

8. F. Diederich, *Cyclophanes* (Royal Society of Chemistry, Cambridge, 1991).
9. C.D. Gutsche, *Calixarenes* (Royal Society of Chemistry, Cambridge, 1991).
10. P.L. Anelli, N. Spencer, and J.F. Stoddart, *J. Am. Chem. Soc.*, **113**, 5131 (1991).
11. P. Kaszynski, A.C. Friedli, and J. Michl, *J. Am. Chem. Soc.*, **114**, 601 (1992).
12. J.S. Lindsey, *New J. Chem.*, **15**, 153 (1991).
13. G.M. Whitesides, J.P. Mathias, and C.T. Seto, *Science*, **254**, 1312 (1991).
14. D.J. Cram, *Nature*, **356**, 29 (1992).
15. P.L. Anelli, P.R. Ashton, R. Ballardini, V. Balzani, M. Delgado, M.T. Gandolfi, T.T. Goodnow, A.E. Kaifer, D. Philp, M. Pietraskiewicz, L. Prodi, M.V. Reddington, A.M.Z. Slawin, N. Spencer, J.F. Stoddart, C. Vicent, and D.J. Williams, *J. Am. Chem. Soc.*, **114**, 193 (1992).
16. *Supramolecular Chemistry*, V. Balzani and L. De Cola, Eds., (Kluwer, Dordrecht, 1992).
17. S. Serroni, G. Denti, S. Campagna, A. Juris, M. Ciano, V. Balzani: *Angew. Chem. Int. Edit. Engl.*, **31**, 1193 (1992).
18. E. Buchlein, W. Wehner, F. Vögtle, *Synthesis*, 155 (1978)
19. For recent papers see, for example, a. M.A. Coffin, M.R. Bryce, A.S. Batsanov, J.A.K. Howard, *J. Chem. Soc. Chem. Commun.*, 552 (1993); b. R. Mülhaupt, C. Wörner, *Angew. Chem. Int. Ed. Engl.*, **32**, 1306 (1993); c. Z. Xu, J.S. Moore, *Angew. Chem. Int. Ed. Engl.*, **32**, 1354; (1993); d. F. Moulines, L. Djakovitch, R. Boese, B. Gloaguen, W. Thiel, J.-L. Fillaut, M.-H. Delville, D. Astruc, *Angew. Chem. Int. Ed. Engl.*, **32**, 1075 (1993); e. G.R. Newkome, C.N. Moorefield, J.M. Keith, G.R. Baker, G.H. Escamilla, *Angew. Chem. Int. Ed. Engl.*, **33**, 666 (1994); f. C.J Hawker, K.L. Wooley, J.M.J. Fréchet, *J. Chem. Soc. Chem. Commun.*, 925 (1994); g. T. Nagasaki, O. Kimura, M. Ukon, S. Arimori, I. Hamachi, S. Shinkai, *J. Chem. Soc. Perkin Trans.* 1, 75 (1994).
20. For reviews, see: a. D.A. Tomalia, A.M. Naylor, W.A. Goddard III, *Angew. Chem. Int. Ed. Engl.*, **29**, 138 (1990); b. J.M.J. Fréchet, *Science*, **263**, 1710 (1994).
21. A. Juris, V. Balzani, F. Barigelletti, S. Campagna, P. Belser, A. von Zelewsky, *Coord. Chem. Rev.*, **84**, 85 (1988).
22. K. Kalyanasundaram, *Photochemistry of Polypyridine and Porphyrin Complexes* (Academic Press, London, 1991).
23. E.M. Kober, J.V. Caspar, B.P. Sullivan, T.J. Meyer, *Inorg. Chem.*, **27**, 4587 (1988).
24. G. Denti, S. Serroni, L. Sabatino, M. Ciano, V. Ricevuto, S. Campagna, *Gazz. Chim. Ital.*, **121**, 37 (1991).
25. L. Della Ciana, D. Dressick, D. Sandrini, M. Maestri, M. Ciano, *Inorg. Chem.*, **29**, 2792 (1990).
26. a. A.A. Vlcek, *Coord. Chem. Rev.*, **43**, 39 (1982). b. M.K. DeArmond and C.M. Carlin, *Coord. Chem. Rev.*, **36**, 325 (1981). c. A.B.P. Lever, *Inorg. Chem.*, **29**, 1271 (1990).
27. V. Balzani, F. Barigelletti, and L. DeCola, *Topics Current Chem.*, **158**, 31 (1990).
28. V. Balzani and V. Carassiti, *Photochemistry of Coordination Compounds* (Academic Press, London, 1970).

29. C.H. Braunstein, A.D. Baker, T.C. Strekas, and H.D. Gafney, *Inorg. Chem.*, **23**, 857 (1984).
30. K.J. Brewer, W.R. Murphy, S.R. Spurlin, and J.D. Petersen, *Inorg. Chem.*, **25**, 882 (1986).
31. S. Campagna, G. Denti, L. Sabatino, S. Serroni, M. Ciano, and V. Balzani, *J. Chem. Soc. Chem. Commun.*, 1500 (1989).
32. G. Denti, S. Campagna, L. Sabatino, S. Serroni, M. Ciano, and V. Balzani, *Inorg. Chem.*, **29**, 4750 (1990).
33. H.A. Goodwin and F. Lions, *J. Am. Chem. Soc.*, **81**, 6415 (1959).
34. F.H. Case and E. Koft, *J. Am. Chem. Soc.*, **81**, 905 (1959).
35. S. Serroni and G. Denti, *Inorg. Chem.*, **31**, 4251 (1992).
36. R.A. Krause, *Inorg. Chim. Acta*, **22**, 209 (1977).
37. P. Belser and A. von Zelewsky, *Helv. Chim. Acta*, **63**, 1675 (1980).
38. K. Kalyanasundaram and Md. K. Nazeeruddin, *Chem. Phys. Lett.*, **158**, 45 (1989).
39. G. Denti, S. Serroni, S. Campagna, V. Ricevuto, A. Juris, M. Ciano, and V. Balzani, *Inorg. Chim. Acta*, **198-200**, 507 (1992).
40. S. Campagna, G. Denti, S. Serroni, M. Ciano, and V. Balzani, *Inorg. Chem.*, **30**, 3728 (1991).
41. G. Denti, S. Campagna, L. Sabatino, S. Serroni, M. Ciano, and V. Balzani, *Inorg. Chim. Acta*, **176**, 175 (1990)
42. Y. Fuchs, S. Lofters, T. Dieter, W. Shi, R. Morgan, T.C. Strekas, H.D. Gafney, and A.D. Baker, *J. Am. Chem. Soc.*, **109**, 2691 (1987).
43. S. Ernst, V. Kasack, and W. Kaim, *Inorg. Chem.*, **27**, 1146 (1988).
44. S. Campagna, G. Denti, L. Sabatino, S. Serroni, M. Ciano, and V. Balzani, *Gazz. Chim. Ital.*, **119**, 415 (1989).
45. G. Denti, S. Serroni, S. Campagna, V. Ricevuto, and V. Balzani, *Inorg. Chim. Acta*, **182**, 127 (1991).
46. G. Denti, S. Campagna, S. Serroni, M. Ciano, and V. Balzani, *J. Am. Chem. Soc.*, **114**, 2944 (1992).
47. W.R. Murphy, Jr., K.J. Brewer, G. Gettliffe, and J.D. Petersen, *Inorg. Chem.*, **28**, 81 (1989).
48. G. Denti, S. Campagna, L. Sabatino, S. Serroni, M. Ciano, and V. Balzani, in *Photochemical Conversion and Storage of Solar Energy*, E. Pellizzetti and M. Schiavello, Eds. (Kluwer, Dordrecht, 1991), p. 27.
49. G. Denti, S. Serroni, S. Campagna, V. Ricevuto, and V. Balzani, *Coord. Chem. Rev.*, **111**, 227 (1991).
50. S. Serroni, G. Denti, S. Campagna, M. Ciano, and V. Balzani, *J. Chem. Soc., Chem. Commun.*, 944 (1991).
51. S. Campagna, G. Denti, S. Serroni, M. Ciano, A. Juris, and V. Balzani, *Inorg. Chem.*, **31**, 2982 (1992).
52. F.H. Kohnke, J.P. Mathias, and J.F. Stoddart, *Angew. Chem. Int. Ed. Engl. Adv. Mater.*, **28**, 1103 (1989).

53. G. Denti, S. Serroni, S. Campagna, A. Juris, and V. Balzani, *Mol. Cryst. Liq. Cryst.*, **234,** 79 (1993).
54. G. Denti, S. Serroni, S. Campagna, A. Juris, M. Ciano, and V. Balzani, in *Perspectives in Coordination Chemistry,* A.F. Williams, C. Floriani, and A. Merbach, Eds., (VCH, Basel, 1992), p. 153.
55. G. Predieri, C. Vignali, G. Denti, and S. Serroni, *Inorg. Chim. Acta,* **205,** 145 (1993).
56. A. Juris, V. Balzani, S. Campagna, G. Denti, S. Serroni, G. Frei, and H.U. Güdel, *Inorg. Chem.*, **33,** 1491 (1994).
57. Charge transfer transitions between remote centers can also be expected, but they give rise to much weaker absorptions; see C.A. Bignozzi, C. Paradisi, S. Roffia, and F. Scandola, *Inorg. Chem.*, **27,** 408, (1988); N.E. Katz, C. Creutz, and N. Sutin, *Inorg. Chem.*, **27,** 1687 (1988).
58. S. Roffia, M. Marcaccio, C. Paradisi, F. Paolucci, V. Balzani, G. Denti, S. Serroni, and S. Campagna, *Inorg. Chem.*, **32,** 3003 (1993).
59. J.B. Flanagan, S. Margel, A.J. Bard, and F.C. Anson, *J. Am. Chem. Soc.*, **100,** 4248 (1978).
60. S. Campagna, V. Ricevuto, G. Denti, S. Serroni, A. Juris, M. Ciano, and V. Balzani, *Syntheses and Methodologies in Inorganic Chemistry,* S. Daolio, M. Fabrizio, P. Guerriero, E. Tondello, and P.A. Vigato, Eds. (University Press, Padova, 1992), p. 128.
61. a. P.C. Bradley, N. Kress, B.A. Hornberger, R.F. Dallinger, and W.H. Woodruff, *J. Am. Chem. Soc.*, **103,** 7441 (1989). b. P.J. Carrol and L.E. Bruss, *J. Am. Chem. Soc.*, **109,** 7613 (1987). c. T. Yabe, D.R. Anderson, L.K. Orman, Y.J. Chang, and J.B. Hopkins, *J. Phys. Chem.*, **93,** 2302 (1989). d. L.F. Cooley, P. Bergquist, and D.F. Kelley, *J. Am. Chem. Soc.*, **112,** 2612 (1990).
62. D.P. Hader and M. Tevini, *General Photobiology* (Pergamon, Oxford, 1987).
63. V. Balzani and F. Scandola, *Supramolecular Photochemistry* (Horwood, Chichester, 1991), ch. 6.
64. C.K. Ryu and R.H. Schmehl, *J. Phys. Chem.*, **93,** 7961 (1989).
65. M. Furue, T. Yoshidzumi, S. Kinoshita, T. Kushida, S. Nozakura, and M. Kamachi, *Bull. Chem. Soc. Jpn.*, **64,** 1632 (1991).
66. F. Scandola, M.T. Indelli, C. Chiorboli, and C.A. Bignozzi, *Top. Curr. Chem.*, **158,** 73 (1990).
67. V. Balzani, S. Campagna, G. Denti, and S. Serroni, in *Photoprocesses in Transition Metal Complexes, Biosystems and Other Molecules: Experiment and Theory,* E. Kochanski, Ed. (Kluwer, Dordrecht, 1992), p. 233.
68. G.A. Crosby, *Acc. Chem. Res.*, **8,** 231 (1975).
69. V. Balzani and F. Scandola, *Supramolecular Photochemistry* (Horwood, Chichester, 1991), ch. 12.
70. V. Balzani, L. Moggi, and F. Scandola, in *Supramolecular Photochemistry,* V. Balzani, Ed. (Reidel, Dordrecht, 1987) p. 1.

4
Molecular Tectonics

JAMES D. WUEST[1]

INTRODUCTION

Nature makes extensive use of hydrogen bonds to regulate the association of biological molecules. The strength and directionality of hydrogen bonds make them particularly well suited for promoting the reversible organization of molecules in geometrically predictable ways. As a result, nature is able to use them to help accomplish a wide variety of critical and subtle tasks, including the recognition and binding of selected molecules, the catalysis of key reactions, the storage, replication, and expression of genetic information, and the production of organized structural materials. The remarkable diversity and ingenuity of these natural processes challenge chemists to devise unnatural compounds that use hydrogen bonds to perform similar tasks or to invent entirely new applications that take full advantage of the characteristic features of hydrogen bonds.

Our plan for exploiting hydrogen bonds in new ways is a simple one, yet the results promise to have broad practical and theoretical significance. We have begun to design and synthesize molecules that incorporate characteristic, well-defined patterns of hydrogen-bond donors and acceptors.[2] These molecules are "sticky"; that is, their ability to participate in extensive hydrogen bonding endows them with a strong tendency to associate and to form complex aggregates in solution and in the solid state.[3] If the arrays of hydrogen-bonding sites are cleverly oriented, the resulting supramolecular aggregation will tend to generate predictable and adjustable architectures. Since these structures are built from individual sticky subunits, we call the subunits *tectons,* from the Greek word, τεκτων, for "builder," and we define *molecular tectonics* as the science and art of supramolecular construction using tectonic subunits.[2b] These special names are justified because the high strength and predictable directionality of the intertectonic attractions sharply distinguish them from the weaker and more diffuse forces typically involved in other processes of self-assembly, including crystallization. In effect, the strategy of molecular tectonics promises to give chemists the elements of a molecular-scale construction set that can be used to build an infinite variety of predictably ordered structures. A principal advantage of this strategy is that the supramolecular aggregates are formed spontaneously by the self-association of sticky tectonic subunits, and the tedious bond-by-bond syntheses normally employed to build complex structures can be avoided.

The individual tectons from which more complex aggregates are constructed must be able to participate in the formation of unusually strong, specific, and reliable hydrogen-bonding motifs. In principle, this will allow hydrogen bonding to override other intermolecular interactions and thereby control aggregation of the tectonic subunits. A particularly attractive motif is the cyclic hydrogen-bonding array defined by dimers of 2-pyridone and related heterocycles (eq. 1).[4-6] This motif is attractive

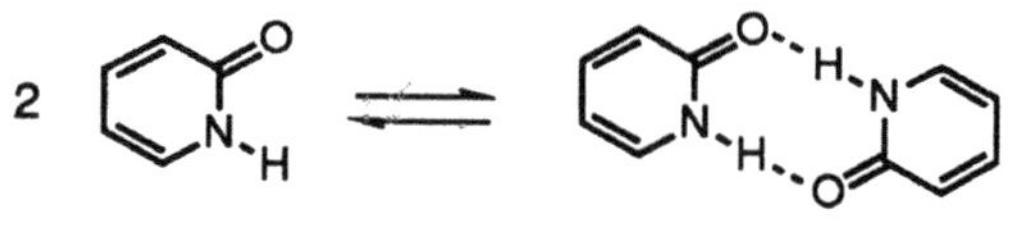

(1)

because it is strong, well studied, and dependable; in addition, pyridones are chemically robust and easy to synthesize. Although they exist in equilibrium with their hydroxypyridine tautomers, the more polar pyridone form normally predominates in solution and in the solid state.[4,5] Ultrasonic studies indicate that the thermodynamic parameters for eq. (1) are $\Delta G° = -2.7$ kcal/mol, $\Delta H° = -5.9$ kcal/mol, and $\Delta S° = -10.7$ eu in $CHCl_3$ at 25°C.[6] Under these conditions, 2-pyridone is approximately 80% dimeric at 10^{-1} M and 2% dimeric even at 10^{-4} M.

AMPLIFICATION AND REGULATION OF INTERTECTONIC ADHESION

Although the hydrogen-bonding motif defined by dimers of 2-pyridone and related heterocycles is strong, it nevertheless contains only two hydrogen bonds. For this reason, the incorporation of a single pyridone ring cannot be expected to guarantee that a tecton will aggregate in a specific way, since two hydrogen bonds may not be able to override other interactions that may favor different modes of association. Fortunately, the strength of the basic motif can be amplified easily by using spacers to link two or more 2-pyridones together. This is an important idea, since it means that we do not need to make entirely new types of tectons in order to achieve stronger intermolecular attractions. Tectons prepared by linking 2-pyridones together will have a larger number of opportunities for forming hydrogen bonds, and the precise mode of tectonic aggregation will depend critically on the particular pattern of hydrogen-bond donors and acceptors that is created. For example, if two pyridones are connected in series at the 3- and 6-positions by a rigid spacer that prevents intramolecular hydrogen bonding, the resulting dipyridone (**1**) is self-complementary and can form an antiparallel duplex held together by four hydrogen bonds (eq. 2). In contrast, isomeric 6,6′-linked dipyridone **2** and the analogous 3,3′-linked structure are not self-complementary and will be forced to form linear oligomers (eq. 3).

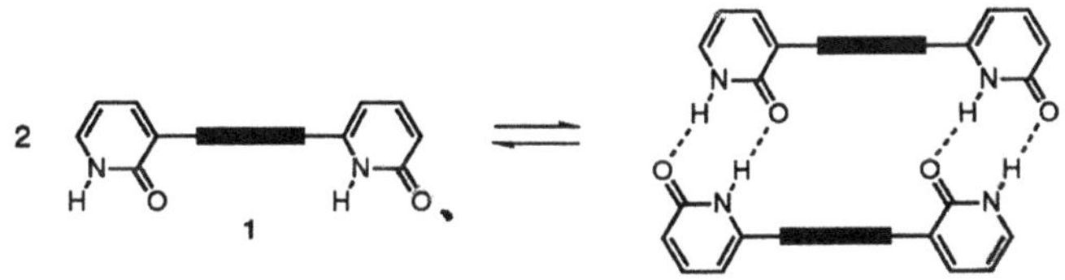

2 **1** ⇌ (2)

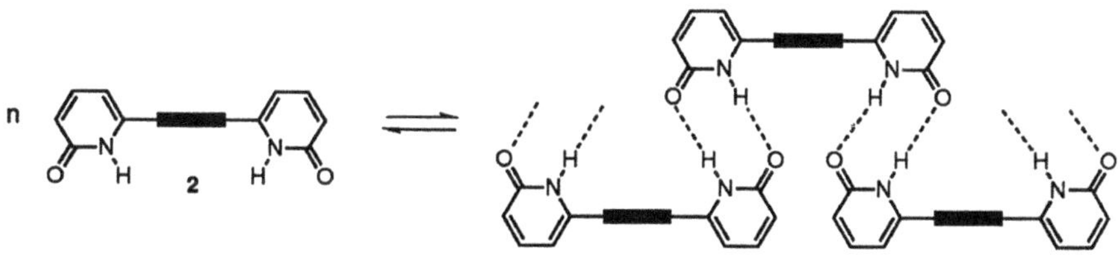

n **2** ⇌ (3)

We have tested this hypothesis by synthesizing rigid acetylenic dipyridones **3** and **4** and by studying their associative behavior.[2d] The syntheses are summarized in Scheme 4-1. Treatment of 2,6-dibromopyridine with $PhCH_2OH$ and KOH[7] provided 2-bromo-6-(phenylmethoxy)pyridine (**5**), which was coupled[8] with acetylene to give the symmetric acetylenic dipyridine **6**. Deprotection using hot CF_3COOH[9] then yielded the corresponding dipyridone **3**. The synthesis of isomeric dipyridone **4** was longer but equally straightforward. Coupling[10] of 2-bromo-6-(phenylmethoxy)pyridine (**5**) with (trimethylsilyl)acetylene provided compound **7**,

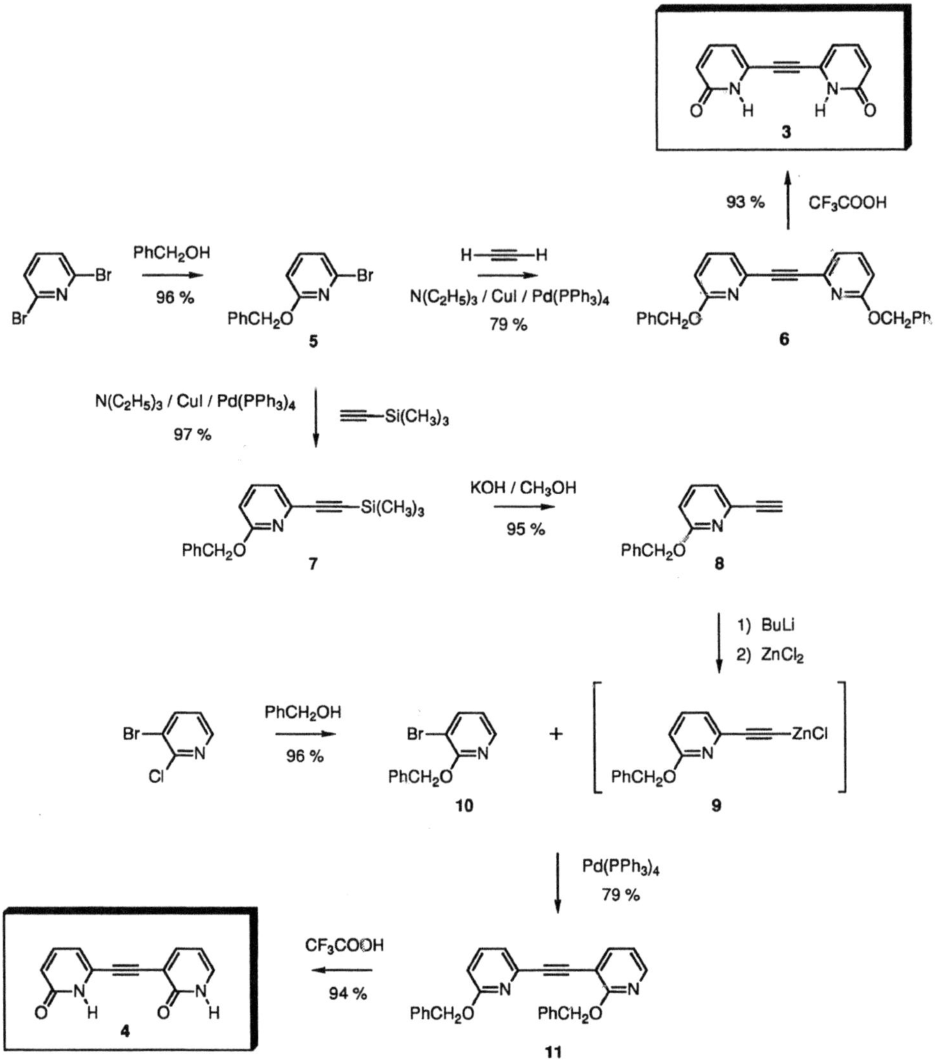

SCHEME 4-1

which was converted into pyridylacetylene **8** by standard deprotection using methanolic KOH. Deprotonation and treatment with $ZnCl_2$ then gave the intermediate zinc acetylide **9**. Coupling[11] with 3-bromo-2-(phenylmethoxy)pyridine (**10**), which was derived from 3-bromo-2-chloropyridine[12] by standard methods,[7] then yielded acetylenic dipyridine **11**. Normal deprotection[9] finally provided the corresponding self-complementary dipyridone **4**. The simplicity of these syntheses illustrates the advantage of our strategy for making molecules with multiple sites of hydrogen bonding by linking 2-pyridones together.

Vapor-pressure osmometric studies confirmed that acetylenic dipyridones **3** and **4** both exist largely as solvated monomers in CH_3OH and other strongly hydrogen-bonding polar solvents. In $CHCl_3$, however, both compounds are strongly associated. Because of their inherently low solubility, we could not determine their association constants by using the standard procedure of variable-concentration ^1H NMR spectroscopy to follow changes in the chemical shifts of hydrogen atoms participating in hydrogen bonds. Instead, we turned to infrared (IR) spectroscopy and compared the ratios of absorbances near 1672 cm^{-1} and 1654 cm^{-1}, which correspond to monomeric and dimeric pyridones, respectively. As expected, the associative behavior of non-self-complementary dipyridone **3** resembles that of 2-pyridone itself. In contrast, self-complementary dipyridone **4** is almost exclusively dimeric at 25 °C even at the lowest concentrations that could be studied conveniently (3.6×10^{-4} M). This observation establishes that $-\Delta G°$ for the self-association of dipyridone **4** in $CHCl_3$ at 25° must be at least 6.5 kcal/mol.

Such strong self-association can occur only if dipyridone **4** forms an antiparallel duplex held together by four hydrogen bonds. It is possible to estimate that $\Delta H°$ for the formation of this duplex will be approximately -11.8 kcal/mol, twice $\Delta H°$ for the dimerization of 2-pyridone itself. Because of the rigidity of dipyridone **4**, $\Delta S°$ for its self-association should be similar to $\Delta S°$ for the dimerization of 2-pyridone.[13] We can therefore estimate that $-\Delta G°$ for the self-association of dipyridone **4** to form a duplex should be approximately 8.6 kcal/mol in $CHCl_3$ at 25 °C. The similarity of this estimate and the experimentally determined minimum value helps confirm that dipyridone **4** exists in solution primarily in the form of an antiparallel duplex held together by four hydrogen bonds. The rigid self-complementary structure of dipyridone **4** ensures that each pyridone subunit makes an approximately additive contribution to the overall enthalpy of dimerization.[14]

These conclusions are supported by the results of X-ray crystallographic studies of dipyridones **3** and **4**. As expected, non-self-complementary isomer **3** associates in the solid state to form the polymeric motif shown in Figure 4-1, while self-complementary isomer **4** exists as the discrete antiparallel duplex shown in Figure 4-2. Curiously, the duplex is distinctly nonplanar; in each dipyridone subunit, the average planes of the pyridone rings make an angle of 29°. This suggests that the hydrogen bonds in the duplex are not ideal,[15] and that they accept small deformations from a perfect linear geometry in response to packing

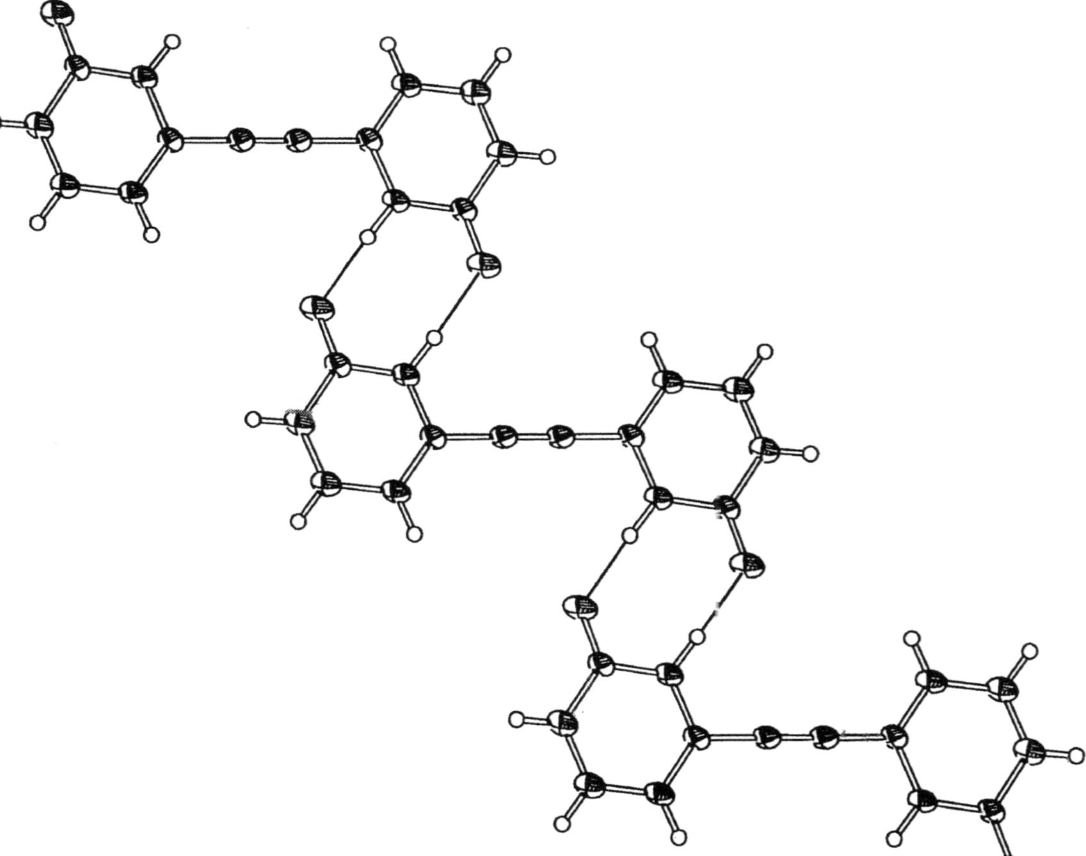

FIGURE 4-1 ORTEP drawing of the structure of the rigid non-self-complementary dipyridone 3 showing three tectonic subunits and part of the network of hydrogen bonds. Non-hydrogen atoms are represented by ellipsoids corresponding to 50% probability, and hydrogen atoms are shown as spheres of arbitrary size. Hydrogen bonds are represented by narrow lines.

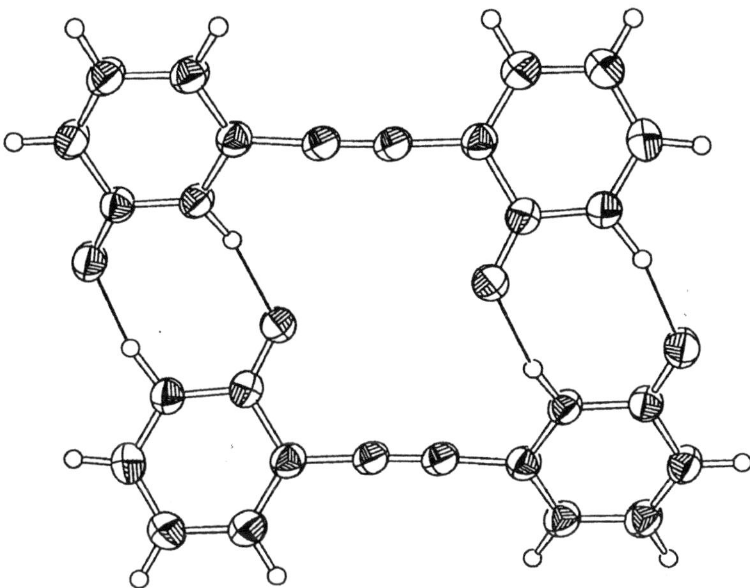

FIGURE 4-2 ORTEP drawing of the dimeric structure of the rigid self-complementary dipyridone **4**. Non-hydrogen atoms are represented by ellipsoids corresponding to 50% probability, and hydrogen atoms are shown as spheres of arbitrary size. Hydrogen bonds are represented by narrow lines.

forces or in order to minimize the repulsion of the two carbonyl oxygen atoms directed toward the interior of the duplex. However, this deviation from an ideal geometry is not costly enough energetically to prevent the duplex from being formed or to keep the contribution of individual pyridone units to the overall enthalpy of dimerization from being approximately additive.

A wide variety of tectons with similar associative properties can be constructed by using different rigid spacers to join two pyridones together. For example, compounds **12** and **13**,[16] which are close relatives of self-complementary dipyridone **4**, have similarly strong tendencies to

12 **13**

form duplexes by self-association. Moreover, the pyridone rings can be replaced by sticky sites derived from other heterocycles with similar properties of self-association. Derivatives of 4-pyrimidinone are attractive substitutes, for example, since 4-pyrimidinone itself exists in the form of dimer **14** in the solid state.[17] In addition, substituted derivatives are relatively inexpensive, and their π-electron-deficient aromatic

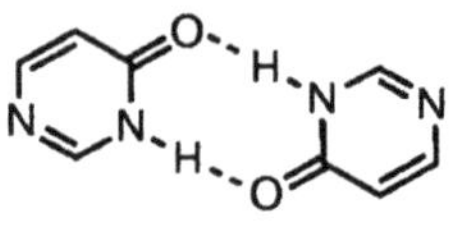

14 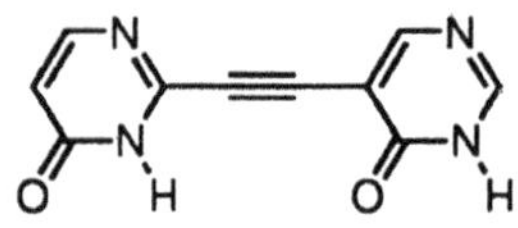 **15**

rings make them particularly susceptible to nucleophilic attacks and coupling reactions of the type described in Scheme 4-1.[18] We have taken advantage of these special features by synthesizing acetylenic dipyrimidinone **15** and related self-complementary analogues of dipyridone **4.**[19] These examples demonstrate that a wide variety of molecules with predictable associative properties can be created by the simple expedient of using rigid spacers to link sticky heterocyclic subunits.

By simple extension, this strategy can generate self-complementary tectons even more strongly self-associated than dipyridones **4, 12,** and **13** or dipyrimidinone **15.** This can be achieved by linking larger numbers of sticky heterocyclic subunits in series. For example, we have synthesized the rigid self-complementary tripyridone **16** by successively coupling 3-ethynyl-2-(phenylmethoxy)pyridine[2a] and the isomeric 2-ethynyl-6-(phenylmethoxy)pyridine (**8,** Scheme 4-1) with 3,6-dibromo-2-(phenylmethoxy)pyridine (**17**),[20] followed by standard deprotection with

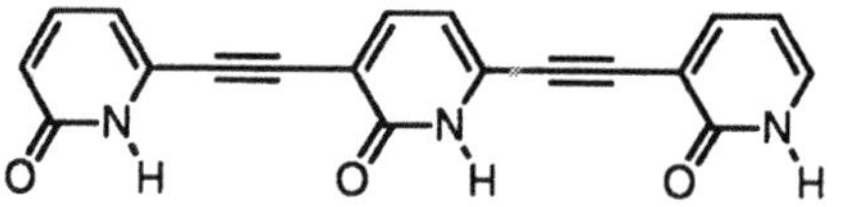

16 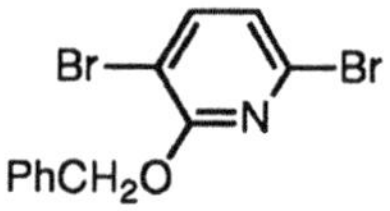 **17**

hot CF_3COOH.[9] Dibromopyridine **17** could be prepared by treating 2,3,5,6-tetrabromopyridine[21] with $PhCH_2OH$ and KOH,[7] followed by regioselective halogen-metal exchange (BuLi, −100 °C) and protonation. Subsequent coupling reactions of compound **17** occur with a high degree of regioselectivity because the 2- and 6-positions in pyridines are typically more activated toward substitution than the 3- and 5-positions.[22]

Tripyridone **16** is designed to form an antiparallel duplex held together by six nearly ideal hydrogen bonds. Unfortunately, we have not yet been able to obtain crystals that would allow us to determine its structure and to confirm that it forms a duplex in the solid state. Furthermore, its very low solubility in organic solvents has prevented us from using osmometric or spectroscopic methods to study its association in solution. Nevertheless, the rigid and almost perfectly self-complementary structure of tripyridone **16** allows us to estimate that $\Delta H°$ for the formation of a duplex will be approximately -17.7 kcal/mol in $CHCl_3$, three times $\Delta H°$ for the dimerization of 2-pyridone itself, and that $\Delta S°$ will be similar to the value observed for 2-pyridone.[13] We therefore expect that $-\Delta G°$ for the self-association of tripyridone **16** in $CHCl_3$ at 25 °C will be approximately 14.5 kcal/mol. Further extension of the chain of sticky sites will increase this value by approximately 5.9 kcal/mol each time another 2-pyridone is added, without changing $\Delta S°$ substantially.[13] It is therefore possible to imagine the straightforward construction of self-complementary tectons that can form duplexes held together by hydrogen bonds with a collective force that surpasses the strength of covalent bonds. Our simple strategy of amplification by systematically increasing the number of participating hydrogen bonds therefore allows us to devise tectons that produce aggregates with any desired degree of cohesion and structural integrity. In principle, we can bridge the gap that distinguishes weak complexes joined reversibly by a small number of hydrogen bonds from well-defined compounds joined essentially irreversibly by covalent bonds.

Another significant advantage of our strategy for devising molecules with a strong tendency to associate is that simple modifications of the spacer or the site of linkage can have dramatic effects on the nature of the aggregate. For example, acetylenic dipyridone **18** is very similar to self-complementary isomer **4**, but the pyridone rings in compound **18** are joined at posi-

3 **18** ⇌ **19** (4)

tions 3 and 5 rather than at positions 3 and 6. As a result, the array of hydrogen-bonding sites in dipyridone **18** is no longer self-complementary in a duplex but becomes perfectly self-complementary only in the triplex **19** (eq. 4). We have prepared dipyridone **18** and the analogous dipyrimidinone **20** by standard coupling procedures,[23] and we are now studying their association in solution and in the solid state.[24]

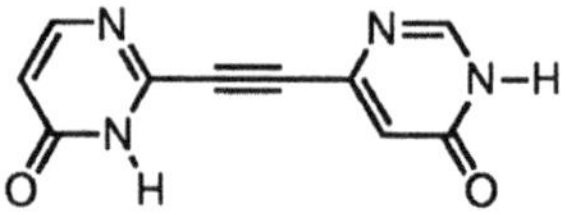

20

The high degree of rigidity of dipyridone **18** allows us to estimate that ΔH° for the formation of triplex **19** will be approximately −17.7 kcal/mol in $CHCl_3$, three times ΔH° for the dimerization of 2-pyridone, and that ΔS° will be approximately −21.4 eu, twice ΔS° for the dimerization of 2-pyridone.[13] As a result, ΔG° for the cyclic trimerization of dipyridone **18** should be about −11.3 kcal/mol in $CHCl_3$ at 25 °C, making triplex **19** the dominant species even in relatively dilute solutions.

AGGREGATION OF TECTONS WITH FLEXIBLE SPACERS

In general, the strongest and most predictable self-association will be achieved when tectons incorporate sticky sites joined by rigid spacers, but impressive degrees of aggregation can nevertheless be attained even when the spacers are modestly flexible. For example, vapor-pressure osmometry and ^1H NMR spectroscopy have been used to show that self-

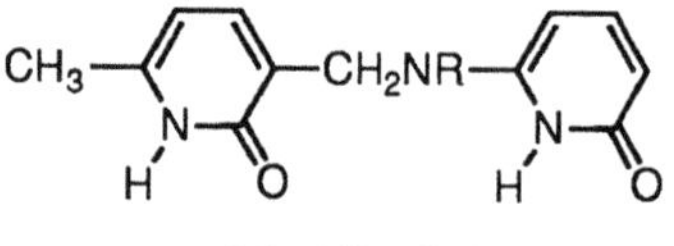

21 ($R = C_2H_5$)

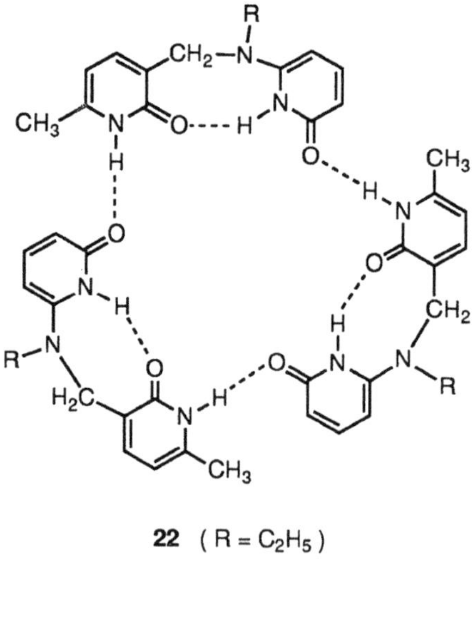

22 ($R = C_2H_5$)

complementary dipyridone **21**, which incorporates a flexible aminomethyl spacer, exists primarily as an antiparallel duplex in $CHCl_3$ at 25 °C even at concentrations as low as 1.6×10^{-3} M.[2c] The dimer cannot be ideal, since constraints imposed by the spacer and by the presence of an intramolecular hydrogen bond disfavor conformations that accommodate normal linear intermolecular hydrogen bonds.[15] As a result, solutions of dipyridone **21** also contain significant amounts of higher aggregates, possibly including the cyclic trimer **22** with intermolecular hydrogen bonds that are more nearly linear. Furthermore, dipyridone **21** does not exist as a discrete duplex in the solid state but forms instead a polymeric chain with both intramolecular and intermolecular hydrogen bonds (Figure 4-3). In general, the behavior of dipyridone **21** suggests that tectons with flexible structural elements will not be capable of the strongest self-association, nor will the resulting aggregates have three-dimensional structures that are easy to predict.

CHARACTERIZATION OF TECTONIC AGGREGATES IN SOLUTION

Vapor-pressure osmometry, routine variable-temperature or variable-concentration ^{1}H NMR spectroscopy, IR spectroscopy, and other standard techniques can be used effectively to determine the degree and nature of tectonic aggregation in solution. In addition, we have recently found that NMR spectroscopy at very low temperatures can be a powerful tool for direct thermodynamic, kinetic, and structural studies of tectonic aggregation directed by multiple hydrogen bonds.[25] This is possible because the rates of association and dissociation are retarded entropically and enthalpically, respectively, since multiple hydrogen bonds must be formed or cleaved. As a result, association and dissociation are no longer diffusion-controlled. For this reason, it is possible in favorable circumstances to see separate signals characteristic of monomeric tectons and various higher aggregates when NMR spectra are recorded at suitably low temperatures. For example, hydroxypyridine monomer **23**[26] and pyridone dimer **24** can be distinguished

2 **23** (R = C_2H_5) ⇌ **24** (R = C_2H_5) (5)

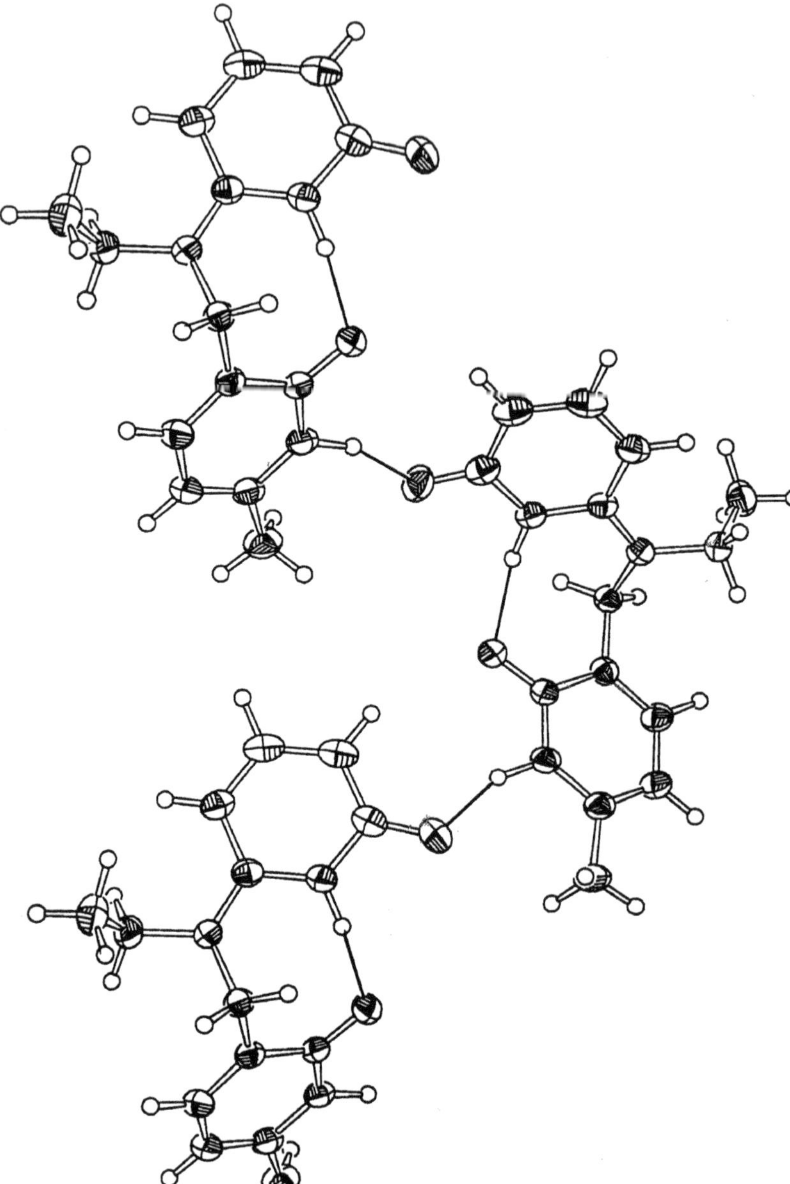

FIGURE 4-3 ORTEP drawing of the structure of the flexible self-complementary dipyridone **21** showing three tectonic subunits and part of the network of hydrogen bonds. Non-hydrogen atoms are represented by ellipsoids corresponding to 40% probability, and hydrogen atoms are shown as spheres of arbitrary size. Hydrogen bonds are represented by narrow lines.

in ^1H and ^{13}C NMR spectra at −118 °C in (THF-d_8) at 0.093 M. By changing the temperature and concentration, we can determine $\Delta H°$ and $\Delta S°$ for the equilibrium of eq. (5), as well as the rates of interconversion of monomer **23** and dimer **24.** In addition, NOE studies can be used to make detailed structural assignments for each species present in solution. We are optimistic that this technique will prove to be extremely useful for analyzing more complex examples of tectonic aggregation.

NETWORKS GENERATED BY TECTONIC AGGREGATION

The sticky molecules discussed so far have interesting structures and striking properties of self-association and aggregation, but they still resemble the simple heterocyclic subunits from which they are constructed. It is not obvious that they offer the chemist any dramatic new opportunities. For this reason, use of the new terms *tecton* and *molecular tectonics* to identify molecules of this type and to define the study of their aggregation could

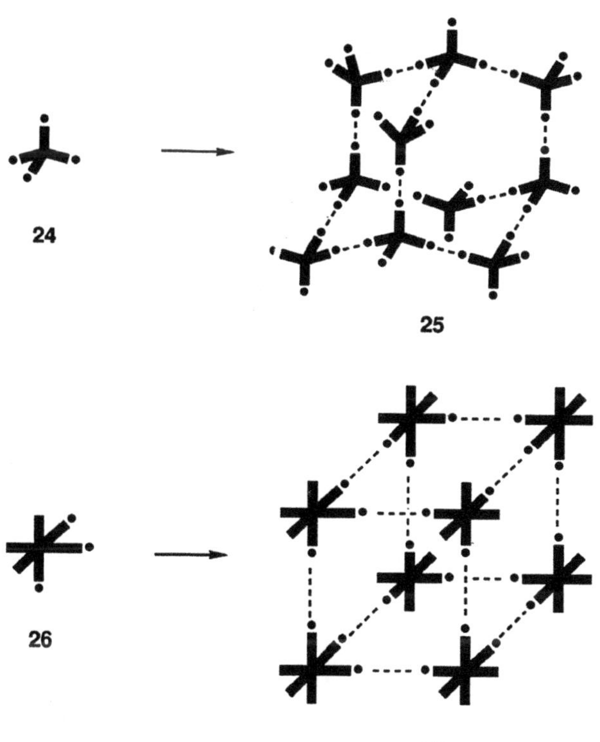

reasonably be considered premature, unnecessary, or pretentious. However, when more complex tectons are created by attaching multiple sticky sites to large, rigid frameworks, the power, elegance, and utility of the concepts of molecular tectonics become evident. For example, we can imagine a tecton **24** created by using rigid arms to connect four sticky sites (•) to a tetrahedral core. Because of its well-defined tetrahedral geometry and the presence of four sticky sites, tecton **24** should be forced to self-associate to form the infinite cubic diamondoid network **25** or a related lonsdaleite lattice. In principle, other infinite repeating lattices can be generated by the self-association of tectons with different core geometries. Furthermore, we believe that discrete aggregates composed of large but finite numbers of subunits can also be created by tectonic association, both in the solid state and in solution. For example, an octahedral inorganic tecton **26** with three facial sticky sites should generate the finite cubic network **27**. In addition to homotectonic aggregates derived from a single self-complementary tecton, heterotectonic aggregates can be formed by the mutual association of different tectonic subunits. These examples suggest that countless opportunities for using molecular tectonics to create novel structures exist for chemists with imagination and the skills required to synthesize appropriate tectonic subunits.

We have recently demonstrated that this strategy can in fact be used to make complex, predictable three-dimensional structures by the spon-

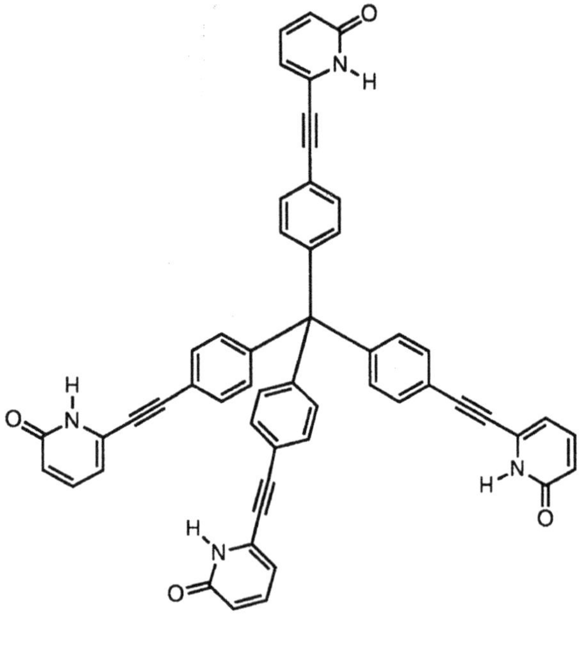

28

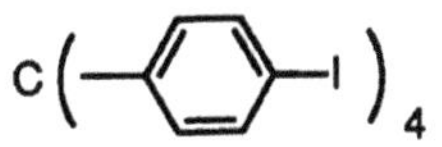

$C(-C_6H_4-I)_4$

29

$C(-C_6H_4-C{\equiv}C-C_5H_3N-OCH_2Ph)_4$

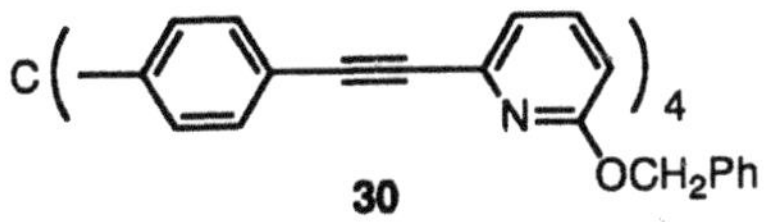

30

taneous self-association of cleverly designed tectonic subunits.[2b,27-29] Tecton **28**, which incorporates four sticky pyridone sites attached rigidly to a tetrahedral core derived from tetraphenylmethane, is designed to generate an infinite diamondoid network by self-association. Tecton **28** was synthesized from tetraphenylmethane[30] in three simple steps. Iodination (I_2, $PhI(OOCCF_3)_2$)[31] gave a 63% yield of tetraiodide **29**, which was then coupled with 2-ethynyl-6-(phenylmethoxy)pyridine (**8**, Scheme 4-1) in the normal manner[10] to provide tetrapyridine **30** in 44% yield. Deprotection with hot CF_3COOH[9] then gave tecton **28** in 100% yield.

After extensive experimentation, we found that tecton **28** could be crystallized from mixed solvents containing significant amounts of carboxylic acids. Use of acetic or propionic acid in hexane or in mixtures of hexane and CH_3OH consistently produced high yields of extremely long needles of composition **28** • 8 RCOOH (R = CH_3, C_2H_5). The morphology of these crystals was surprising, since we did not expect an essentially isotropic diamondoid lattice **25** to permit exaggerated growth in a single direction. In fact, a diamondoid network is not formed under these conditions of crystallization. Instead, an X-ray crystallographic study of **28** • 8 CH_3CH_2COOH revealed that self-assembly of a diamondoid network is thwarted by association of each of the four sticky pyridone sites with two molecules of carboxylic acid, which produces adduct **31**.

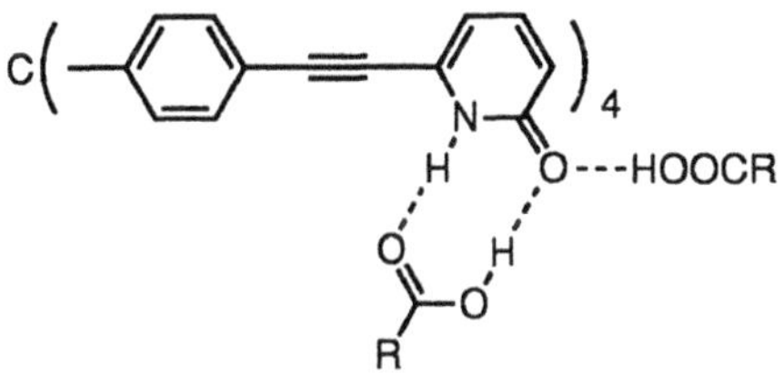

O---HOOCR

31 (R = C_2H_5)

This was unexpected and disappointing, since the strong tendency of pyridones and related heterocycles to self-associate had served as a reliable guiding principle in the successful design of simpler tectons. We were therefore tempted to treat the undesired mode of crystallization of tecton **28** from acetic and propionic acids as an abnormality, and we examined the structure of the predicted diamondoid network **25** in order to understand the origins of the abnormal crystallization. A conspicuous feature of this hypothetical network is the large amount of space not occupied by the tectonic subunits. This is energetically problematic, since it requires that the structure forego the weak, diffuse, but numerous stabilizing interactions that constitute the forces of close packing in exchange for the stronger, more specific, but less numerous forces provided by the network of intertectonic hydrogen bonds. If the network of hydrogen bonds is strong enough, this exchange may be thermodynamically feasible. If not, the open structure will collapse, multiple diamondoid networks will interpenetrate, or the spaces defined by the network will be forced to accommodate molecules of solvent or other guests. Since acetic and propionic acids are relatively small molecules, a correspondingly large number will be needed to fill the voids in an open diamondoid network. The failure of tecton **28** to form the expected network in acetic and propionic acids can therefore be attributed, at least in part, to an especially unfavorable entropy of enclathration. For this reason, we turned to larger carboxylic acids. Crystallization of tecton **28** from butyric acid/hexane/CH_3OH reproducibly gave high yields of prisms of approximate composition **28** • 2 $CH_3CH_2CH_2COOH$. The dramatic change in the morphology and composition of the crystals provided immediate evidence that a completely different structure had been formed. An X-ray crystallographic study of this new structure confirmed that the sticky pyridone sites interact in the expected way to induce self-assembly of the remarkable diamondoid network shown in Figure 4-4.

Since the tetrahedral centers of adjoining tectons are separated by 19-20 Å, the network defines an open, porous structure with enormous chambers and interconnecting passageways. This permits a 7-fold interpenetration of independent diamondoid networks, but significant internal spaces remain open. These spaces enclathrate only butyric acid, even though crystallization occurs in a mixed solvent. We attribute this selectivity to the characteristic size, shape, polarity, and hydrogen-bonding ability of butyric acid, which appear to make it more complementary than competing molecules of hexane or CH_3OH to the interior and faces of the growing crystals. The interstitial guests are surprisingly well ordered and form two parallel columns in four-sided channels aligned with the *b* axis. These particular channels are clearly defined in Figure 4-5,

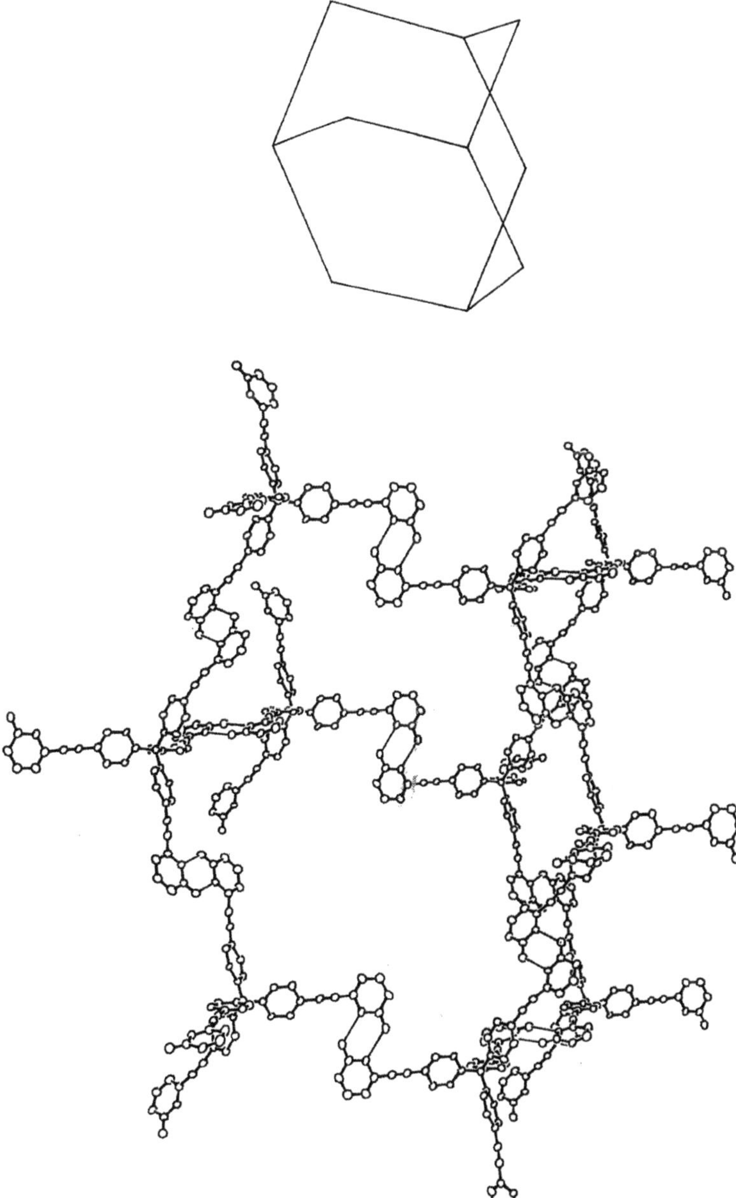

FIGURE 4-4 ORTEP drawing of part of the infinite diamondoid network present in crystals of **28** • 2 $CH_3CH_2CH_2COOH$. The tetrahedral centers of the ten tectons in the drawing define a distorted adamantane, shown at the right at one-half scale. Non-hydrogen atoms are represented by ellipsoids corresponding to 50% probability. Enclathrated butyric acid and all hydrogen atoms have been omitted for clarity. Hydrogen bonds are represented by narrow lines.

which represents a view along the *b* axis of an idealized diamondoid network generated by the self-association of hypothetical tectonic subunits **24.** Figure 4-6 presents a similar view of the actual structure and shows a cross section of four adjacent channels, each containing paired columns of enclathrated butyric acid.

Crystallization of tecton **28** from valeric acid/hexane/CH_3OH provided prisms of approximate composition **28** • 1 $CH_3CH_2CH_2CH_2COOH$ in 76% yield. Again, an X-ray crystallographic study confirmed that self-assembly occurs to give a closely similar diamondoid network, and again

FIGURE 4-5 View along the *b* axis of an idealized diamondoid network generated by the self-association of hypothetical tectonic subunits **24.**

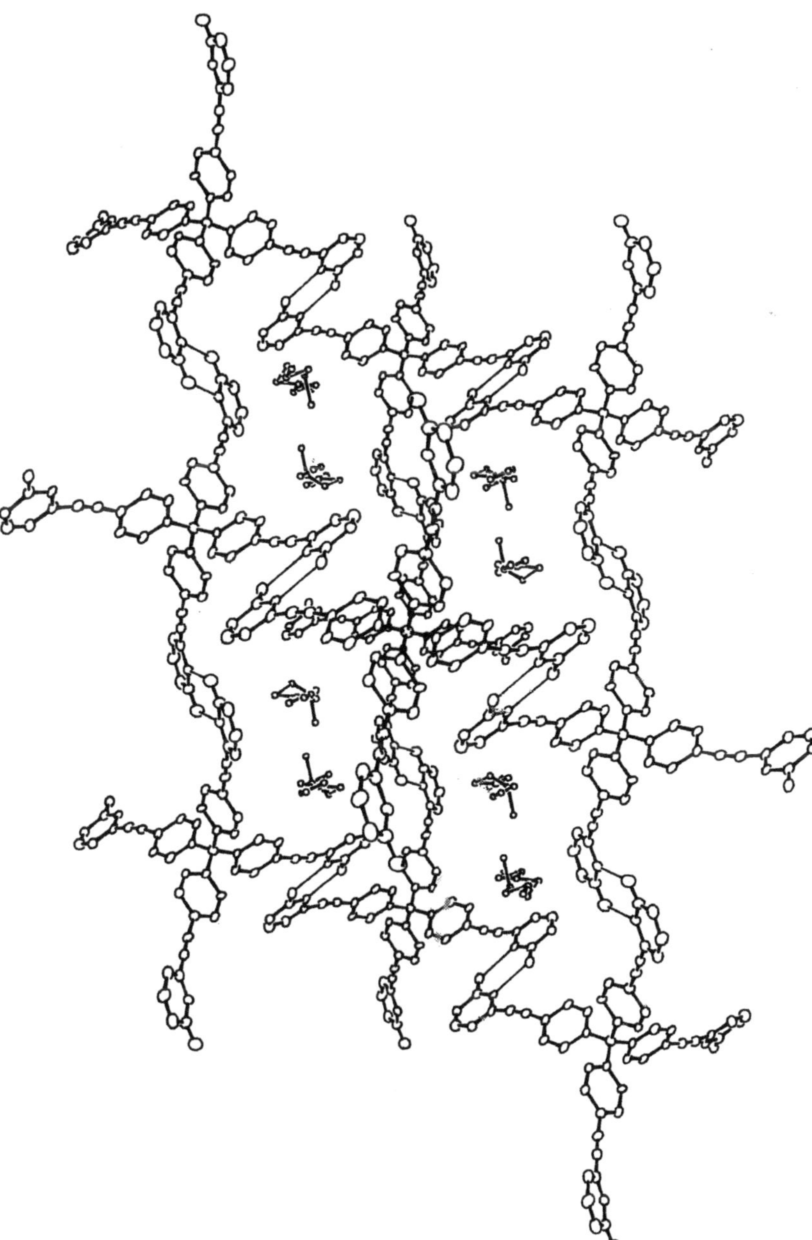

FIGURE 4-6 ORTEP drawing of part of the infinite diamondoid network present in crystals of **28** • 2 $CH_3CH_2CH_2COOH$. The structure is viewed along the *b* axis to provide a clear cross-sectional image of the channels and the paired parallel columns of enclathrated butyric acid. Non-hydrogen atoms are represented by ellipsoids corresponding to 50% probability. All hydrogen atoms have been omitted for clarity. Hydrogen bonds are represented by narrow lines.

only the carboxylic acid is enclathrated. Crystallizations of tecton **28** from isobutyric acid and isovaleric acids gave similar results. In addition, crystallization from a mixed solvent containing propionic and butyric acids in a 1:1 ratio produced a diamondoid network containing largely butyric acid, even though propionic acid by itself favors the alternative mode of crystallization. Together, these observations establish that tecton **28** has an inherently strong tendency to generate porous diamondoid clathrates by self-association when suitable guest molecules are present. This tendency is the result of its rigid tetrahedral geometry and the presence of four sticky sites that ensure a predictable intertectonic adhesion.

OPPORTUNITIES FOR FUTURE RESEARCH IN MOLECULAR TECTONICS

The ability of tetrapyridone **28** to generate a porous diamondoid network by self-association illustrates the power of the concepts of molecular tectonics. It is an encouraging observation, but much work must still be done in order to explore the limitations of molecular tectonics and to establish fully that it is an important general strategy for devising predictably ordered structures. We have begun by synthesizing analogous tectons **32**[32] and **33**[33] with modified tetrahedral geometries, which should

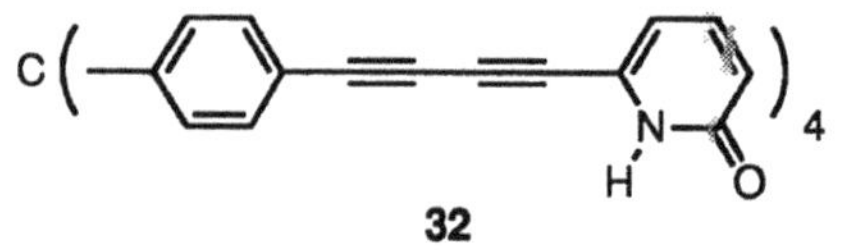

32

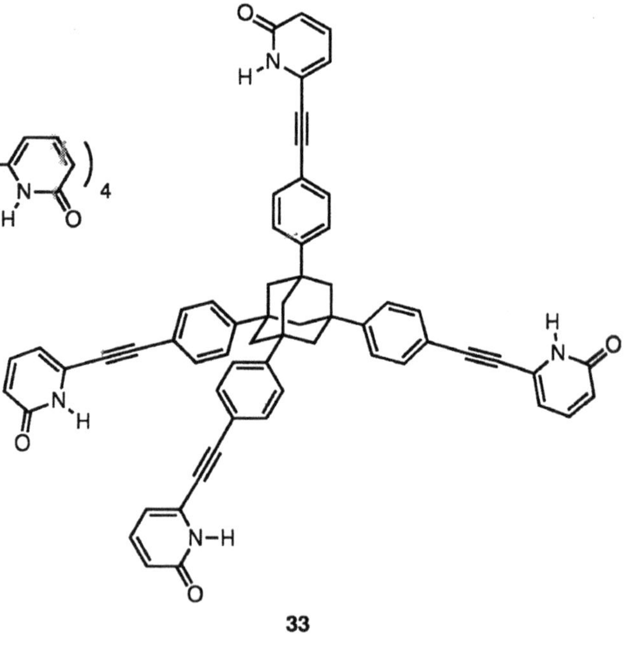

33

self-associate to form diamondoid networks with larger intertectonic separations and internal cavities. In the network derived from compound **32**, for example, the intertectonic distance should increase to 22-23 Å, and the volume of the chambers should increase by approximately 50%. Studies of these tetrahedral tectons and related molecules with different core geometries will establish whether or not the demonstrated ability of tecton **28** to form open, three-dimensional networks is a phenomenon that can be generalized broadly.

Although the self-association of tecton **28** to form a diamondoid network is predictable, the ordering of the enclathrated guests has more subtle origins. Molecules of enclathrated butyric acid in the parallel columns form hydrogen-bonded dimers typical of carboxylic acids. This creates a head-to-head, tail-to-tail orientation of successive molecules in each column. No hydrogen bonds are formed between the two columns, or between the columns and the walls of the channels, so movement of the columns within the channels along the *b* axis should occur relatively freely. Since the columns are retained within a porous host framework by van der Waals forces alone, we were not surprised to observe that loss of butyric acid occurs when the crystals are placed under vacuum at 25 °C. During loss of the guests, the crystals retain their characteristic morphology, but X-ray diffraction experiments have confirmed that the ordered diamondoid network does not remain intact. The framework may be robust enough to withstand a partial loss of enclathrated guests, but a more extensive loss presumably triggers structural changes that arise because the partially empty framework cannot resist forces favoring close packing.

Since the stickiness of the sites that create the hydrogen-bonded network can easily be amplified by using larger numbers of heterocycles connected in series, we are optimistic that we can create modified frameworks strong enough to resist substantial forces favoring close packing. For example, we have synthesized the exceptionally sticky octapyridone **34**,[32] although we have not yet been able to crystallize it. The large

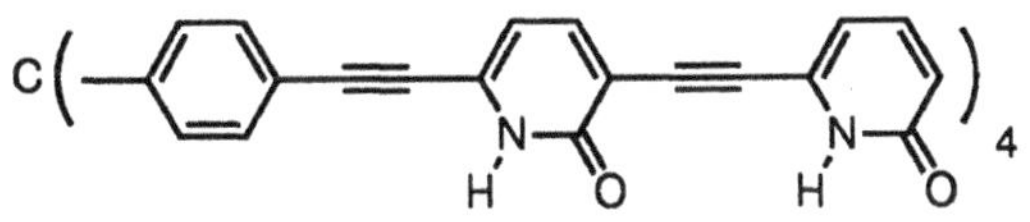

C(N H O N H O)4

34

enthalpy of association characteristic of the multiple sticky sites in tecton **34** and related molecules should provide a powerful driving force for the self-assembly of predictable networks. The exceptionally strong, specific intertectonic attraction caused by hydrogen bonding is designed to overwhelm other intermolecular effects and ensure that solvent molecules and forces favoring close packing do not conspire to disrupt the ordered network. It is therefore possible that frameworks derived from compound **34** and related tectons may be durable enough to resist the powerful forces of close packing even when most enclathrated molecules have been removed by heating under vacuum.

This is a very exciting possibility, since it would produce an ordered, open, and partially empty organic network with some of the useful properties of zeolites and other rigid, inorganic frameworks with micropores. Specifically, materials derived from unusually sticky tectons may be able to act as organic molecular sieves and to be able to include, exclude, and separate molecules according to size, shape, and chemical behavior. Furthermore, substituted tectons may be able to generate networks with cavities bearing functional groups that can modify the selectivity and geometry of enclathration or catalyze reactions of enclathrated guests. Although materials derived from organic tectons will have less thermal stability than zeolites and related inorganic solids, purely organic networks offer compensating advantages. For example, enclathrated molecules can be recovered easily from tectonic aggregates by adding acid or base to cleave the surrounding network of hydrogen bonds. This disassembly should be reversible, since neutralization should regenerate the monomeric tecton and allow its spontaneous self-association. In addition, functionalization of networks derived from organic tectons is simple, structural modifications are easy to achieve, and substituents designed to have specific chemical effects can be introduced logically. Since the structure and functionalization of organic networks can in principle be modified at will, tectonic aggregates offer extraordinary opportunities for using a single material to effect separation and catalysis. We are optimistic that these ambitious goals are realistic, and that chemists can use the strategy of molecular tectonics to create a wide variety of novel structures with specific architectural and functional features.

Acknowledgment

I am grateful to the students and collaborators whose names are cited in the references for their enthusiastic participation in this project and their many creative suggestions. Our work was supported financially by grants from the Natural Sciences and Engineering Research Council of Canada, the Ministère de l'Éducation du Québec, the Canada Council, and Imperial Oil Limited.

NOTES AND REFERENCES

1. Killam Research Fellow, 1992-1994. This chapter was submitted in June, 1992.

2. a. F. Persico and J.D. Wuest, *J. Org. Chem.*, **58**, 95 (1993). b. M. Simard, D. Su, and J.D. Wuest, *J. Am. Chem. Soc.*, **113**, 4696 (1991). c. M. Gallant, M.T. Phan Viet, and J.D. Wuest, *J. Org. Chem.*, **56**, 2284 (1991). d. Y. Ducharme and J.D. Wuest, *ibid.*, **53**, 5787 (1988).

3. For recent references to related studies of the use of hydrogen bonds to control molecular aggregation, see J.A. Zerkowski, C.T. Seto, and G.M. Whitesides, *J. Am. Chem. Soc.*, **114**, 5473 (1992). T.K. Park, Q. Feng, and J. Rebek, Jr., *ibid.*, **114**, 4529 (1992). J.-M. Lehn, M. Mascal, A. DeCian, and J. Fischer, *J. Chem. Soc., Perkin Trans. 2*, 461 (1992). J.-D. van Loon, R.G. Janssen, W. Verboom, and D.N. Reinhoudt, *Tetrahedron Lett.*, **33**, 5125 (1992). J.S. Nowick, N.A. Powell, E.J. Martinez, E.M. Smith, and G. Noronha, *J. Org. Chem.* **57**, 3763 (1992). T. Kato, H. Adachi, A. Fujishima, and J.M.J. Fréchet, *Chem. Lett.* 265 (1992). C. Hilger, M. Dräger, and R. Stadler, *Macromolecules*, **25**, 2498 (1992). F. Garcia-Tellado, S.J. Geib, S. Goswami, and A.D. Hamilton, *J. Am. Chem. Soc.*, **113**, 9265 (1991). T.R. Kelly, G.J. Bridger, and C. Zhao, *ibid.*, **112**, 8024 (1990). X. Zhao, Y.-L. Chang, F.W. Fowler, and J.W. Lauher, *ibid.*, **112**, 6627 (1990). M.C. Etter, *Acc. Chem. Res.*, **23**, 120 (1990).

4. For crystallographic studies of the association of simple pyridones, see M. Nethaji and V. Pattabi, *Acta Crystallogr., Sect. C*, **C45**, 509 (1989). U. Oth, H. Guth, E. Hellner, H. Dannöhl, and A. Schweig, *Z. Kristallogr.*, **169**, 185 (1984). J.N. Low and C.C. Wilson, *Acta Crystallogr., Sect. C*, **C39**, 1688 (1983). M.M. Kadooka, M.Y. Chang, H. Fukami, P.J. Scheuer, J. Clardy, B.A. Solheim, and J.P. Springer, *Tetrahedron*, **32**, 919 (1976). Å. Kvick and S.S. Booles, *Acta Crystallogr., Sect. B*, **B28**, 3405 (1972). J. Almlöf, Å. Kvick, and I. Olovsson, *ibid.*, **B27**, 1201 (1971). B.R. Penfold, *Acta Crystallogr.*, **6**, 591 (1953).

5. M. Kuzuya, A. Noguchi, and T. Okuda, *J. Chem. Soc., Perkin Trans. 2*, 1423 (1985). A. Fujimoto, K. Inuzuka, and R. Shiba, *Bull. Chem. Soc. Jpn.*, **54**, 2802 (1981). M. Chevrier, O. Bensaude, J. Guillerez, and J.-E. Dubois, *Tetrahedron Lett.*, **21**, 3359 (1980). P. Beak, J.B. Covington, S.G. Smith, J.M. White, and J.M. Zeigler, *J. Org. Chem.*, **45**, 1354 (1980). P. Beak, *Acc. Chem. Res.*, **10**, 186 (1977). K.-Å. Engdahl and P. Ahlberg, *J. Chem. Res., Synop.*, 340 (1977). J. Elguero, C. Marzin, A.R. Katritzky, and P. Linda, *Adv. Heterocycl. Chem., Suppl.*, **1**, 1 (1976). M.H. Krackov, C.M. Lee, and H.G. Mautner, *J. Am. Chem. Soc.*, **87**, 892 (1965).

6. G.G. Hammes and A.C. Park, *J. Am. Chem. Soc.*, **91**, 956 (1969).

7. A.J. Serio Duggan, E.J.J. Grabowski, and W.K. Russ, *Synthesis*, 573 (1980).

8. K. Sonogasniri, Y. Tohda, and N. Hagihara, *Tetrahedron Lett.*, 4467 (1975).

9. J.P. Marsh, Jr., and L. Goodman, *J. Org. Chem.*, **30**, 2491 (1965).

10. T. Sakamoto, M. Shiraiwa, Y. Kondo, and H. Yamanaka, *Synthesis,* 312 (1983).

11. E.-i. Negishi, F.-T. Luo, R. Frisbee, and H. Matsushita, *Heterocycles,* **18,** 117 (1982). A.O. King, E.-i. Negishi, F.J. Villani, Jr., and A. Silveira, Jr., *J. Org. Chem.,* **43,** 358 (1978).

12. H.J. Den Hertog and N.A.I.M Boelrijk, *Recl. Trav. Chim. Pays-Bas,* **70,** 578 (1951).

13. a. This is a crude approximation that neglects the loss of entropy of rotation around triple bonds as well as other factors. b. For a discussion of $\Delta S°$ in the formation of hydrogen-bonded dimers, see A.J. Doig and D.H. Williams, *J. Am. Chem. Soc.,* **114,** 338 (1992). D.H. Williams, *Aldrichimica Acta,* **25,** 9 (1992).

14. Secondary electrostatic interactions in the duplex appear to be similar to those in the dimer of 2-pyridone itself. J. Pranata, S.G. Wierschke, and W.L. Jorgensen, *J. Am. Chem. Soc.,* **113,** 2810 (1991).

15. For a discussion of the ideal geometry of similar hydrogen bonds, see R. Taylor, O. Kennard, and W. Versichel, *Acta Crystallogr., Sect. B,* **B40,** 280 (1984). R. Taylor and O. Kennard, *Acc. Chem. Res.,* **17,** 320 (1984).

16. M. Simard, X. Wang, and J.D. Wuest, unpublished results.

17. M. Simard, L. Vaillancourt, and J.D. Wuest, unpublished results.

18. J. Solberg and K. Undheim, *Acta Chem. Scand.,* **43,** 62 (1989).

19. L. Vaillancourt and J.D. Wuest, unpublished results.

20. L. Richard and J.D. Wuest, unpublished results.

21. T.K. Chen and W.T. Flowers, *J. Chem. Soc., Chem. Commun.,* 1139 (1980).

22. B.C. Uff, In *Comprehensive Heterocyclic Chemistry,* Vol. 2, A.R. Katritzky and C.W. Rees, Eds., (Pergamon Press, Oxford, 1984), p. 315.

23. E. Boucher and J.D. Wuest, unpublished results.

24. For a related study, see S.C. Zimmerman and B.F. Duerr, *J. Org. Chem.,* **57,** 2215 (1992).

25. M. Gallant, M.T. Phan Viet, and J.D. Wuest, *J. Am. Chem. Soc.,* **113,** 721 (1991).

26. Although 2-pyridones are normally more stable than their hydroxypyridine tautomers in condensed phases,[4,5] the presence of an amino group in the 6-position favors hydroxypyridine tautomers. A. Fujimoto and K. Inuzuka, *Bull. Chem. Soc. Jpn.,* **63,** 2292 (1990). D.G. de Kowalewski, R.H. Contreras, and C. de los Santos, *J. Mol. Struct.,* **213,** 201 (1989). G.B. Barlin and W. Pfleiderer, *J. Chem. Soc., B* 1425 (1971). E.M. Peresleni, L.N. Yakhontov, D.M. Krasnokutskaya, and Y.N. Sheinker, *Dokl. Akad. Nauk SSSR,* **177,** 592 (1967).

27. For related work, see O. Ermer and L. Lindenberg, *Helv. Chim. Acta,* **74,** 825 (1991).

28. For discussions of related hydrogen-bonded clathrates and engineered

crystals, see E. Weber and M. Czugler, *Top. Curr. Chem.*, **149,** 45 (1988). E. Weber, *ibid.*, **140,** 1 (1987). G.R. Desiraju, *Crystal Engineering. The Design of Organic Solids* (Elsevier, New York, 1989).

29. For recent studies of related structures in which predictable order is maintained by coordination to metals, see T. Kitazawa, S.-i. Mishikiori, A. Yamagishi, R. Kuroda, and T. Iwamoto, *J. Chem. Soc., Chem Commun.*, 413 (1992). R.W. Gable, B.F. Hoskins, and R. Robson, *ibid.*, 762 (1990). B.F. Hoskins and R. Robson, *J. Am. Chem. Soc.*, **112,** 1546 (1990).
30. F.A. Neugebauer, H. Fischer, and R. Bernhardt, *Chem. Ber.*, **109,** 2389 (1976).
31. E.B. Merkushev, N.D. Simakhina, and G.M. Koveshnikova, *Synthesis*, 486 (1980).
32. D. Su and J.D. Wuest, unpublished results.
33. X.-X. Du and J.D. Wuest, unpublished results.

5
Supramolecular Assemblies from "Tinkertoy" Rigid-Rod Molecules

JOSEF MICHL

INTRODUCTION

Molecular "Tinkertoys"

The ultimate objective of the work on molecular assembly in my laboratory is the development of a molecule-sized "Tinkertoy" construction kit analogous to the Tinkertoy kit for children.[1,2] It should consist of a set of connectors and a set of rigid-rod molecules available in a multitude of lengths and diameters. They should not absorb light and should be inert thermally, toward reduction or oxidation, toward solvents, and toward most other chemicals These rigid-rod molecules should be provided with terminal axial substituents suitable for covalent attachment to the connectors. The degree of control over the attachment reactions is to be sufficient for the stepwise assembly of both simple and complicated structures. Both permanent and reversible attachment reactions will be needed, as will suitable temporary blocking groups.

The objects that we ultimately wish to assemble are of two kinds:

1. "Free-floating" constructs such as cubes or more complex shapes containing a relatively small number of rods and connectors. These represent a fairly straightforward extension of standard targets of organic synthesis and will be characterizable by the usual tools, such as NMR spectroscopy, single-crystal X-ray analysis, and so on. In many cases, their solubility will be limited by their size. Other investigators are working on constructs of this type as well and have recently published successful syntheses of

several, based on rods and permanent connectors of types other than those of most interest to us. Perhaps the nicest example are the constructs combining aromatic rings and acetylenic units.[3]

2. "Surface-bonded," covalent, extended solid structures, such as scaffoldings and thin layers, periodic and extensive (macroscopic) in up to two dimensions. In the third dimension they are to be microscopic and aperiodic, assembled one story at a time. For these structures, solubility is not an issue. However, they represent an unprecedented synthetic target, and the assembly and characterization techniques are far more demanding. Full control of the assembly procedures would provide access to "designer solids," novel materials with arbitrarily selected, active groups attached to the scaffolding at preselected points in space. Epitaxial growth of such solids on a suitable single-crystal surface appears attractive at first sight but suffers from the probable misfit of the surface lattice spacing and the rod length, which are extremely unlikely to be commensurate.

Strategy

The development of a molecule-size Tinkertoy system naturally falls into three partially overlapping stages. The first is the synthesis of suitable rods, the second is the identification or synthesis of connectors, and the third is the development of assembly techniques.

Tinkertoy Rods. Over the last several years, we have worked on the synthesis of rigid-rod molecules with axial terminal substituents, and we now have three classes of such structures. One is based on carbon:oligomeric bicyclo[1.1.1]pentanes, or [n]*staffanes* for short (**1**),[1,2,4] and hybrid rods containing other types of cages as well,[5] and two are based on boron:oligomeric 10-vertex (**2**)[6] and 12-vertex (**3**)[6] *p*-carboranes (unfunctionalized structures of type **3** were independently developed in another laboratory as well,[7] and the name *carborod* was suggested for them). Interest

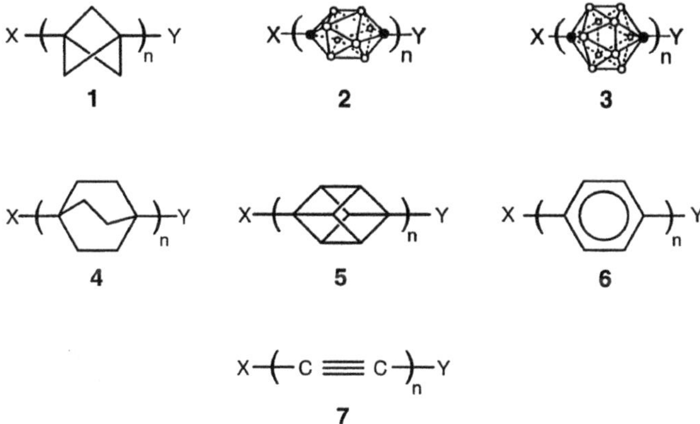

in rod-like molecules is not limited to supramolecular construction, and two other series have been prepared recently in other laboratories: oligomeric bicyclo[2.2.2]octanes (**4**),[8] for which the name [*n*]*rodanes* has been suggested, and oligomeric cubanes (**5**),[9] referred to as *polycubyls*. In these two series of rods, terminally disubstituted derivatives suitable for our purposes are not yet available. One class of rigid, rod-like molecules that has been known for a long time are the *polyphenyls* (**6**), whose symmetry, however, is substantially lower than rod-like; another are the *polyynes* (**7**). Many other, even less symmetrical and more complicated, but still somewhat rod-like molecules are also available, such as various polyimides, or "molecular lines,"[10] but these are typically devoid of truly axial terminal substituents needed for our purposes.

Tinkertoy Connectors and Rod-Connector Assembly. Since an ample supply of various kinds of rod-like molecules is now being developed, offering a selection of various lengths and diameters, we have recently begun to pay attention to the next two stages of the project: the development of connectors and covalent assembly methods for extended solids. The covalent assembly has not yet advanced very far in our hands, and as mentioned above, the best current examples of Tinkertoy-type construction of free-floating objects have come from other laboratories, using less symmetrical, chemically and optically less inert rods, such as DNA[11] or benzene rings and acetylenes.[3]

Two types of connectors are being worked on at present in our laboratory (Figure 5-1). *Point* connectors will have the ability to attach three or more rods by forming new covalent bonds from the connector to the terminal positions on the rods at well-defined angles of less than 180°, resulting in geometries such as trigonal, square, tetrahedral, octahedral, and so on. Metal atoms with multiple ligands are an example of point connectors that most readily comes to mind. The interconnector distance in the final assembly equals the length of the rod used plus the lengths of the new bonds at rod termini.

Star connectors will contain three or more permanently attached, rod-like arms in well-defined directions, leading to trigonal, tetrahedral, or other geometries. Each arm will have the ability to bind covalently, in a linear fashion, to the end of the arm of another connector or to a rod terminus. In this case, two bonds formed during the assembly never originate at the same atom of the connector. Examples of star connectors are adamantanes carrying arms at all four bridgeheads, benzenes carrying arms in positions 1, 3, and 5, and so on. The ultimate node-to-node separation after assembly with star connectors would thus be at least twice the length of the initial arms. If the attachment is of the "star arm

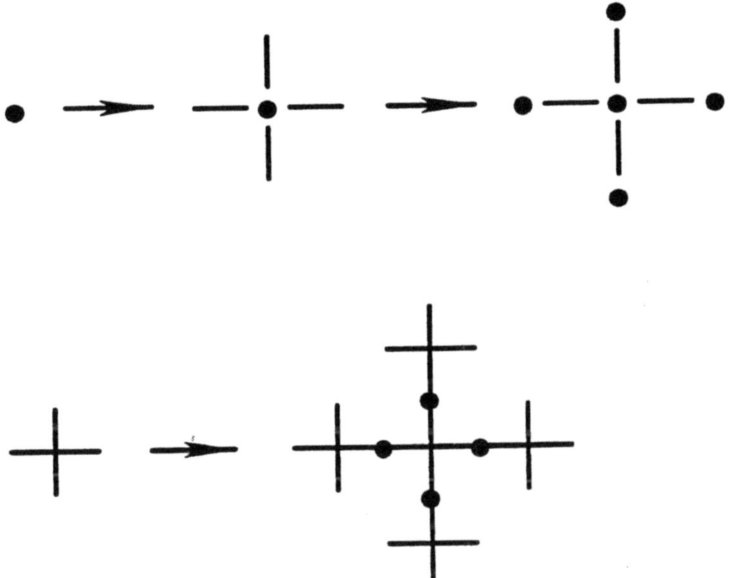

FIGURE 5-1 Schematic of point (top) and star (bottom) connectors.

to rod to star arm" type rather than of the "star arm to star arm" type, the resulting distances between connector centers could be much larger.

Self-Assembled and Langmuir-Blodgett Monolayers. As an intermediate stage in the work toward our ultimate goal, we have decided to examine the utility of our rod-like molecules for supramolecular assembly into monolayers of the self-assembled and Langmuir-Blodgett types. Unlike true designer solids, which will be covalent, these structures rely on van der Waals forces for lateral cohesion and thus should be far less sturdy. Nevertheless, we felt that the preparation of these conventional types of monolayers would represent a useful learning process for the ultimately planned assembly of covalent thin layers.

In this chapter, we survey briefly the synthesis and properties of the [*n*]staffane rods developed in our laboratory. Subsequently, we describe in more detail their use for the production of self-assembled and Langmuir-Blodgett monolayers and the structural characterization of the resulting thin films. Although these films in effect are mere by-products in the development of molecular Tinkertoys, it appears that they will be of some interest in their own right, as rigid rods offer certain advantages over the generally used alkane chains in the assembly of monolayers and multilayers.

[*n*]STAFFANE MOLECULAR RODS

Synthesis

[*n*]Staffanes (**1**) with two terminal substituents are synthesized by radical-induced oligomerization of [1.1.1]propellane with suitable reagents. The initially rather exotic [1.1.1]propellane[12] has become a convenient starting material for synthesis after an efficient procedure for its large-scale preparation was reported.[13] Thus, terminal dithiols (**1**, n = 1-5, X = Y = SH) are accessible from the reaction of [1.1.1]propellane with diacetyldisulfide and separation of the oligomers, followed by hydrolysis[2]:

(AcS)2 hν AcS n SAc HS n SH

Terminal dicarboxylic acids **1**(n = 1-4, X = Y = COOH) are accessible from a similar reaction of [1.1.1]propellane with benzil followed by oxidation[2]:

(PhCO)2 hν PhCO n COPh HOCO n COOH

This process, however, suffers from poor yields, and the first two members of the series are more conveniently prepared otherwise[2,14]:

(MeCO)2 hν MeCO COMe HOCO COOH

1. 0.5 I2 2. hν I I *n*-Bu3SnH (MeCO)2

MeCO COMe HOCO COOH

The first two diacids in the series have been converted into the corresponding dicarbonitriles, diamines, and diisocyanides by standard procedures.[15]

For the purposes of assembly of conventional monolayers, with which we deal in this chapter, singly terminally substituted or differentially terminally disubstituted [*n*]staffanes are required.

Singly terminally substituted [*n*]staffanes are again accessible by radical-induced oligomerization of [1.1.1]propellane with appropriate reaction partners. Insertion into C-H bonds is particularly useful. For instance, the first five terminal monocarboxylic acids have been obtained by separating the oligomers resulting from the addition of [1.1.1]propellane across the CH bond of methyl formate,[2,4] followed by hydrolysis. Higher overall yields result on oxidation of the products of the oligomerizing insertion of [1.1.1]propellane across the α C-H bond of diethyl ether[2]:

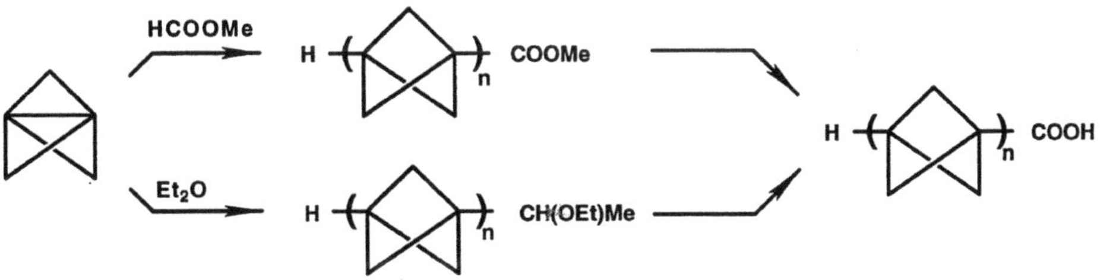

HCOOMe; H–(…)$_n$–COOMe; Et_2O; H–(…)$_n$–CH(OEt)Me; H–(…)$_n$–COOH

Differentially terminally disubstituted [*n*]staffanes are accessible by insertion into other bonds, such as C-I. This procedure has been used to prepare the following series of iodo acids[16]:

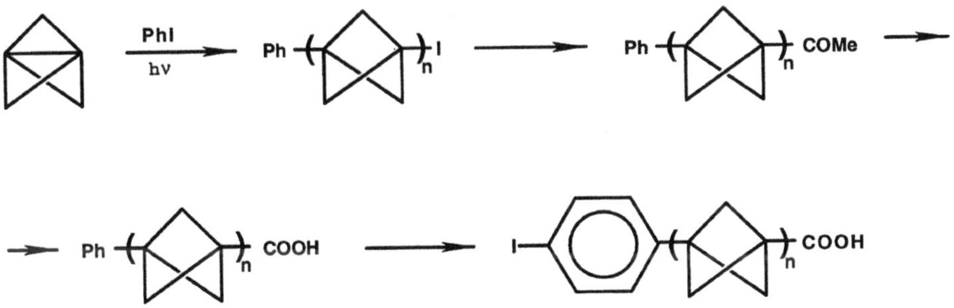

PhI, hν; Ph–(…)$_n$–I; Ph–(…)$_n$–COMe; Ph–(…)$_n$–COOH; I–C$_6$H$_4$–(…)$_n$–COOH

It is also possible to differentiate two equal terminal substituents on an

[*n*]staffane by means of a partial reaction followed by separation. Thus, the above-mentioned terminally disubstituted 3,3$^{(n-1)}$-bis(acetylthio)[*n*]staffanes have been converted to the differentially terminally disubstituted 3-mercapto-3$^{(n-1)}$-acetylthio[*n*]staffanes by partial hydrolysis[2]:

AcS–[staffane]$_n$–SAc ⟶ HS–[staffane]$_n$–SAc

The synthesis of the *p*-carborane-based rods **2**[6] and **3**[6,7] takes advantage of the acidity of the hydrogens of the CH groups of the *p*-carboranes.[17,18] It is based on oxidative coupling of lithium salts with Cu(II), originally reported about 20 years ago[19] but not utilized much until recently. Like those in the parent monomeric *p*-carboranes, the terminal CH groups in their oligomers are acidic. This permits facile terminal functionalization and represents a considerable synthetic advantage over [*n*]staffanes.

An example of utilization of CH acidity for the differentiation of the two termini in the 10-vertex oligocarborane series[6] is the introduction of silyl substituents and their partial removal:

1. BuLi
2. $CuCl_2$
56 %
21 %
92 %
1. BuLi
2. Et_3SiCl
94 %
1. BuLi
2. Et_3SiCl
94 %
1. BuLi
2. Et_3SiCl
Et_3Si
$SiEt_3$
62 %
$[Bu_4N]F$
53 %
$[Bu_4N]F$
52 %
$[Bu_4N]F$
H
$SiEt_3$

Related procedures have been used to prepare disubstituted rods of the 12-vertex oligo-*p*-carborane series with two equal or two different substituents[6]:

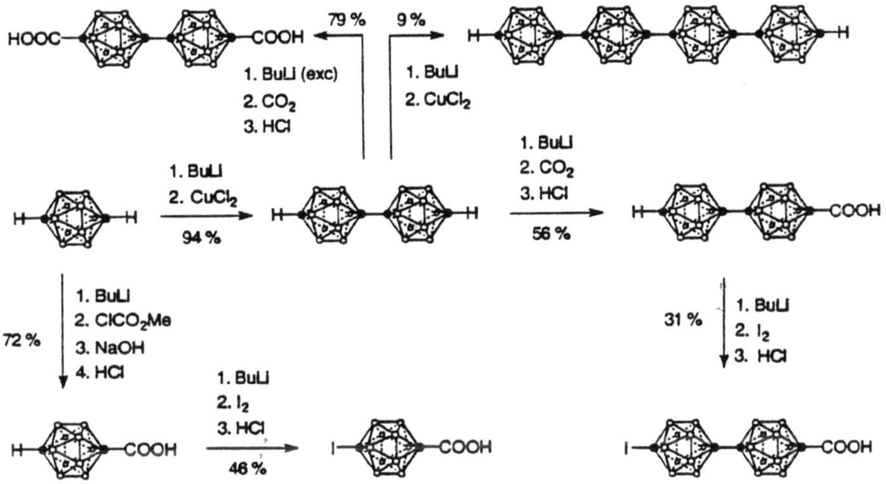

Properties

The availability of several different classes of rod-like molecules offers many advantages. The [*n*]staffane rods are synthesized from cheap, readily available starting materials; in addition, they have the smallest length increments and the smallest diameter, only about 5.5 Å. The diameters of the 10-vertex (7.1 Å) and 12-vertex (7.65 Å) carborane rods are significantly larger, as are the length increments (the intracage C-C distances are 3.4 and 3.1 Å, respectively); the starting materials are expensive at present, but synthetic transformations, particularly terminal functionalization, are facile. The [*n*]staffane rods are stable up to about 300 °C, and the carborane rods are stable well above 500 °C. It is possible to combine two[5] and undoubtedly also a larger number of different structural elements in a single rod; this provides an even finer gradation of available lengths, on the order of 1 Å.

Molecular Structure

[*n*]Staffanes have been available for several years and have been characterized fairly thoroughly, although not yet in the detail that they will ultimately require. In contrast, we have worked with the boron-based rods for a shorter time and have far less information about their properties. In the following, we describe the molecular structure, the vibrational modes, and some aspects of the electronic structure of [*n*]staffanes.

Investigations of about a dozen molecules of type **1** by single-crystal

X-ray diffraction[20,21] showed that they are linear, as expected. An example is shown in Figure 5-2. Each additional bicyclo[1.1.1]pentane cage adds roughly 3.33 Å to the rod's length. This increment represents the sum of the remarkably small, nonbonded, intracage bridgehead-bridgehead C-C distance, about 1.85 Å, and of the equally remarkably short intercage C-C bond length, about 1.46-1.48 Å. The bond length, about 0.1 Å less than expected ordinarily for a hexasubstituted ethane, is accounted for by reference to Bent's rules[22] and to the high degree of *s* character in the bridgehead orbital used in forming the exocyclic bond.

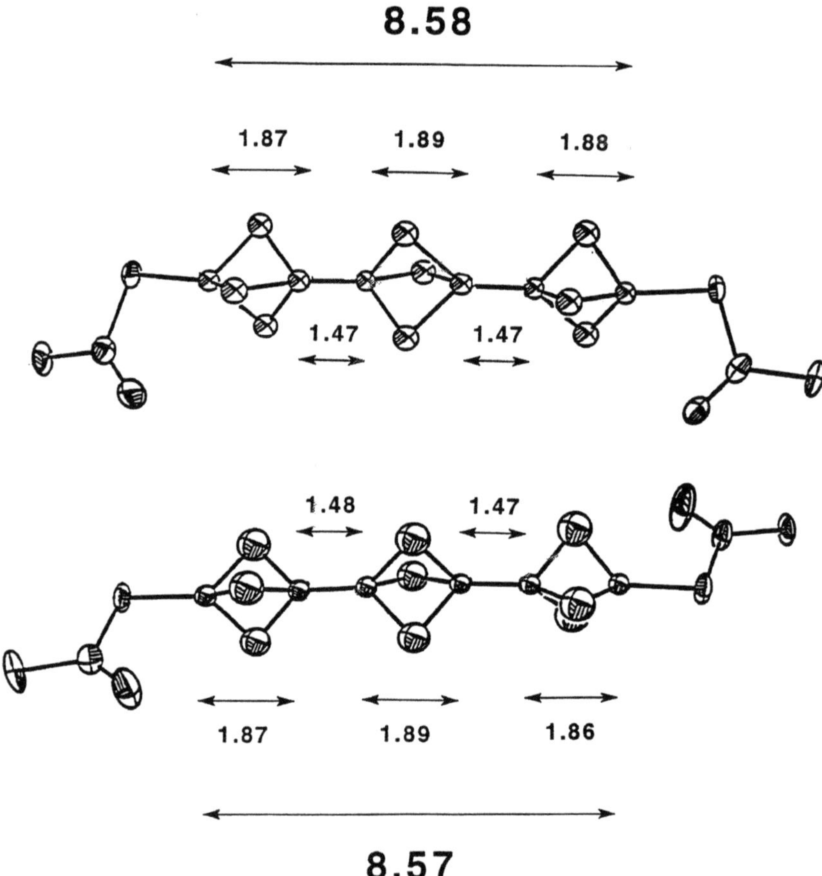

FIGURE 5-2 Single-crystal X-ray structures of two conformers of 3,3″-bis(acetylthio)[3]staffane.[20] Distances in Å.

Vibrations

The vibrations of [*n*]staffanes have been studied in some detail[23] in argon matrix isolation and in stretched polyethylene[24] as solvent. To a reasonable approximation, the vibrational normal modes can be separated into two classes, intercage and intracage. In the intercage modes of vibration, the individual bicyclo[1.1.1]pentane cages do not distort much and mostly move relative to each other. In the intracage modes, the motions are localized within each cage, and there is not much motion of one cage relative to another.

There are four classes of intercage vibrations: accordion, rod-bend, rolling, and internal rotation. The high strength of the short intercage C-C bond contributes to the high frequency of the "accordion" vibration of the [*n*]staffane rod. The situation is actually somewhat complicated: a high-frequency stretching motion of the exocyclic C-C bond couples with a low-frequency skeletal deformation (symmetric C-C stretching) intracage motion that modulates the intracage bridgehead-bridgehead distance. The coupling yields predominantly intercage accordion modes at 1300-1400 cm^{-1} and primarily intracage skeletal deformation modes near 300 cm^{-1}.

The other expected intercage vibrations have not been observed experimentally. The rod-bending modes are calculated to have very low frequencies, all below 200 cm^{-1}, and to be extremely weak both in infrared (IR) and in Raman spectra. In the longer rods, these bends occur as a series of fundamental normal modes of vibration that correspond to a fundamental and the mechanical overtones of vibrations in a macroscopic string, with an increasing number of nodes along the length of the staffane rod. The nodeless vibration has the lowest frequency, calculated at 125 cm^{-1} in [2]staffane, 61 cm^{-1} in [3]staffane, 35 cm^{-1} in [4]staffane, and so on.

The intercage rolling modes are calculated to have somewhat higher frequencies, about 300-400 cm^{-1}, whereas the internal rotations are calculated to have very low frequencies, below 100 cm^{-1}. The calculated barrier to internal rotation is about 2 kcal/mol, quite low for what is fundamentally a hexasubstituted ethane but reasonable in view of the angles at which the six substituents are attached to the ethane carbons.

Although the intercage vibrations are more interesting for the "mechanical" properties of the staffane rods, the intracage vibrations have been more useful so far in that they are intense in the IR spectrum and serve as a diagnostic tool in studies of rod alignment. These are the vibrations that are already present in a single bicyclo[1.1.1]pentane cage. Their numbers grow regularly as the number of such cages in the

[*n*]staffane increases. The most important of these fundamental modes are (1) a very intense in-phase combination of all CH_2 wagging motions, which occurs at 1215 cm^{-1} and is polarized along the long axis of the [*n*]staffane molecule, and (2) the doubly degenerate, out-of-phase combination of symmetric CH_2 stretching motions, polarized perpendicular to the long axis, which occurs as a Fermi doublet near 2880 and 2915 cm^{-1} due to interaction with two quanta of a CH_2 scissors fundamental. A C-H stretching peak near 2980 cm^{-1} is of mixed polarization due to band overlap and is not very useful for diagnostic purposes.

The stretching vibration of the terminal CH group present in monosubstituted [*n*]staffanes is observed near 3000 cm^{-1}. It is polarized along the long molecular axis but is weak relative to the stretching vibrations of the much more numerous CH_2 groups.

Electronic Structure

Inspection of the molecular geometry at the bicyclo[1.1.1]pentane cage bridgeheads shows that the back lobes of the hybrid orbitals used to make the exocyclic C-C bonds must overlap significantly and suggests that there will be significant electronic communication between neighboring cages in [*n*]staffanes. This should have interesting consequences for their electronic structure. In particular, instead of a series of essentially non-interacting C-C σ orbitals located at the molecular axis, the [*n*]staffanes ought to contain a σ-homoconjugated, linear array of hybrid orbitals of σ symmetry with strongly alternating resonance integrals. This should then lead to delocalized σ molecular orbitals and should be reflected in molecular properties.

Unsubstituted [*n*]staffanes do not absorb above 200 nm in the UV, but their ionization potentials are relatively low and converge to about 8.5 eV as the rod gets longer.[2] It is likely but not certain that the highest occupied molecular orbital in these hydrocarbons is of σ symmetry. Relatively strong interactions have been observed[25] between π-symmetry orbitals on substituents attached to the bridgeheads.

A probe that has demonstrated the suspected presence of σ bond delocalization in the [*n*]staffane skeleton was the EPR spectroscopy of bridgehead [*n*]staff-3-yl radicals.[26] These formally fully localized radicals exhibit easily detectable long-range hyperfine coupling to bridgehead protons in positions as distant as ζ and ι. An example is shown in Figure 5-3.

In view of the large π- and σ-symmetry interactions, one might expect electron transfer through the [*n*]staffane rods to be unusually efficient. The high symmetry of the [*n*]staffane rods permits a separate investi-

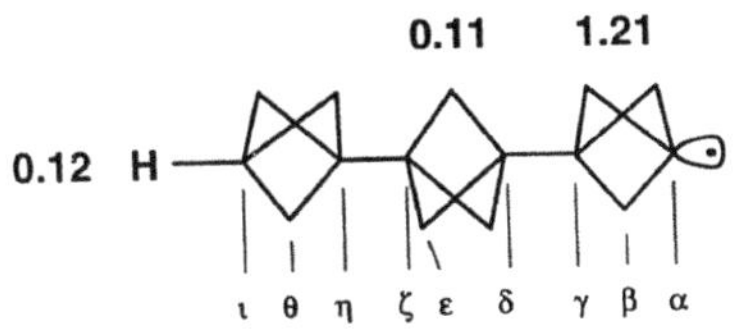

FIGURE 5-3 Proton hyperfine coupling constants in the [3]staff-3-yl radical.[26]

gation of the transfer of charge as a function of the symmetry of the orbitals involved, provided that the donor and acceptor moieties are of sufficiently high symmetry themselves, but such an investigation has not yet been carried out. The only measurements performed to date used the low-symmetry $(NH_3)_5RuS$- substituents on both rod ends. These do not permit a separation of charge transmission through σ- and π-symmetry orbitals. From measurements of the charge-transfer bands in mixed-valence complexes, it was concluded[27] that the electron transmission rate is rather high, $1 \times 10^7\ s^{-1}$ for $n = 1$ and $2 \times 10^6\ s^{-1}$ for $n = 2$, comparable to those observed previously for spiro linked cyclobutanes.[28] Additional work is clearly required.

[*n*]STAFFANE ROD ASSEMBLIES

Langmuir-Blodgett Monolayers and Multilayers

So far, these have been prepared from various salts of [*n*]staffane-3-carboxylic acids[29] (**1**, X=H, Y= COOH) and of the iodo acid derived from dimeric 12-vertex *p*-carborane[6] (**3**, X=I, Y=COOH). Here, we shall describe only the results for the former, about which much more is known.

[*n*]Staffane-3-carboxylic acids and their salts are too water-soluble if *n* equals 1 or 2, and they are too insoluble in suitable organic solvents if *n* equals 5 or higher. Monolayers have been prepared successfully using chloroform and chloroform-acetonitrile solutions of the [*n*]staffane-3-carboxylic acids with *n* equal to 3, using a concentrated solution of $CaCl_2$ as the aqueous phase, and with *n* equal to 4, using pure water or any of a large number of dilute salt solutions. The transfer onto solid substrates was facile for $n = 4$ and demanding for $n = 3$. Examples of pressure-area isotherms for the two cases are shown in Figure 5-4 and demonstrate that the molecular surface area extrapolated to zero pressure, about 26.0 $Å^2$ per molecule, is the same regardless of the rod length. This suggests that the nature of rod packing in the monolayer is the same in the two

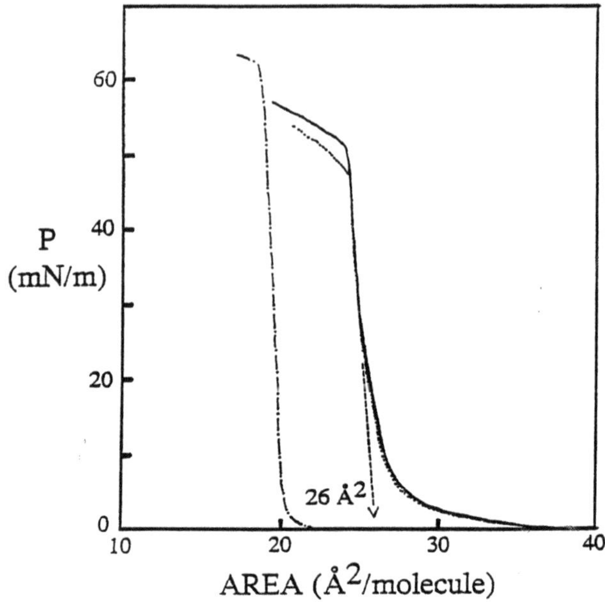

FIGURE 5-4 Langmuir-Blodgett isotherms for Ca [*n*]staffane-3-carboxylates with *n* = 3 (dotted line) and *n* = 4 (solid line). Cadmium arachidate (dot-dash line) is shown as a reference. (Reproduced by permission from Ref. 29)

cases. Results obtained with a variety of counterions are nearly identical. The films collapse on further compression to an area of about 24.5 $Å^2$ per molecule; those obtained with divalent cations, except for Mg^{2+}, collapse at about 50 mN/m, and those obtained with monovalent cations or Mg^{2+} at about half that pressure. They appear to be quite crystalline, since the pressure remains constant or even increases further on continued compression, suggesting that large film segments slide over each other. Both at zero pressure and at collapse, the area per molecule is considerably larger than the corresponding areas observed for the standard LB films produced from cadmium arachidate, also shown in Figure 5-4 for comparison (20 and 18 $Å^2$, respectively).

This result is compatible with the relatively large diameter found in the single-crystal X-ray structures of the [*n*]staffane rods[20,21] compared with an *n*-alkane chain. In the crystal structures of the parent [*n*]staffane hydrocarbons (Figure 5-5),[21] the rods are packed in a herringbone-type fashion, with axes tilted at about 20° from the crystal axis, and mesh so that the bulging cages of one rod snuggle against the narrower waists between the cages on a neighboring rod. The perpendicular distance between the rod axes on the closest neighbors is about 5.4 Å.

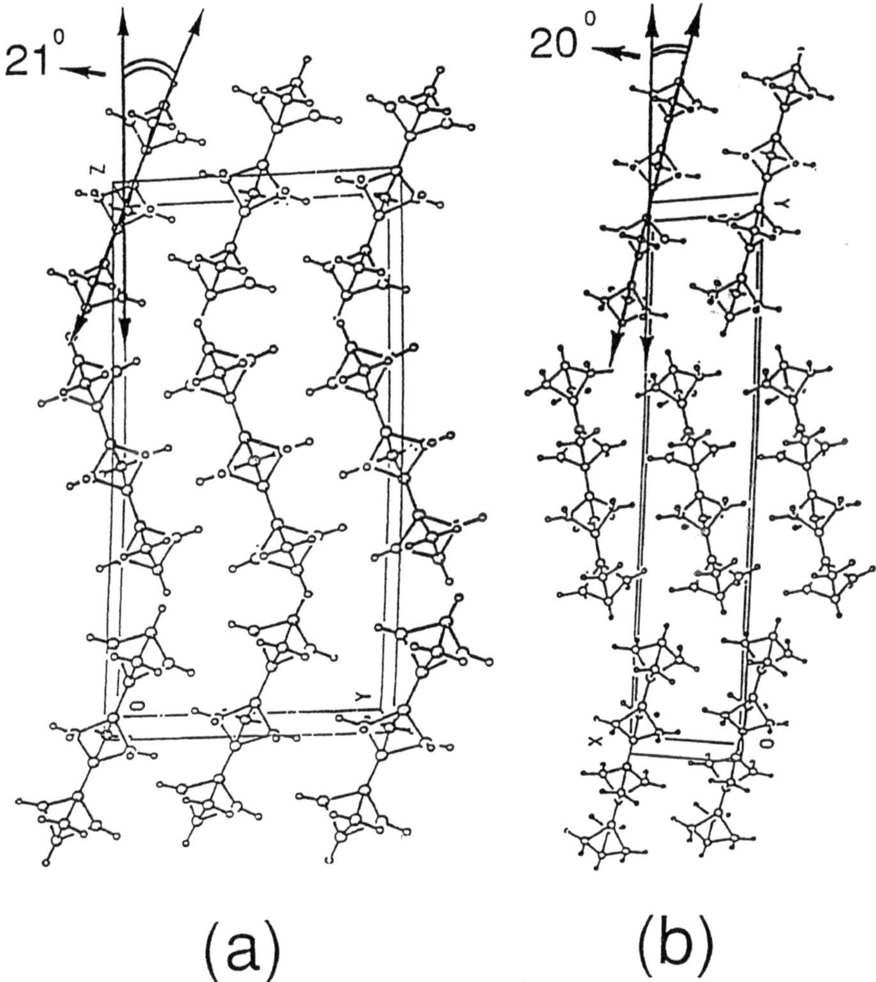

FIGURE 5-5 Single-crystal X-ray packing diagrams for (a) [3]staffane projected along the crystal x axis and (b) [4]staffane projected along the z axis. (Reproduced by permission from Ref. 21)

If the [n]staffane rods in the Langmuir-Blodgett film were perpendicular to the water surface, such meshing would result in a radius of only about 2.7 Å. Half of the molecules would have to be 1.67 Å farther from the water surface than the rest, and the projection of each rod on the water surface would be only about 23 Å^2. The observed area per molecule would depend on the nature of the two-dimensional packing arrangement of the projected circles, which dictates how much of the water surface remains uncovered in a compactly packed film. The choices of packing are limited by the requirement of perfect meshing of each rod with all of its neighbors. For instance, in a square array, which permits such a meshing, the area per molecule would be about 29 Å.[2]

In the absence of all meshing, with all molecules perpendicular to the surface, the perpendicular orientation would cause each molecule to project a circle with an area close to 33 Å^2, and the observed area per molecule in a tightly packed film would have to be larger. If the rods were tilted, the observed area per molecule would have to be larger still. Since the observed area is only 26 Å^2 at zero pressure and 24.5 Å^2 at 50 mN/m, some meshing clearly takes place. If the meshing is achieved by keeping all molecules equidistant from the water surface and tilting them all in the same direction by about 20° from the surface normal, as in the single crystals of [*n*]staffanes (Figure 5-5), the observed area would be about 28 Å^2 per molecule. Clearly, the area per molecule alone is insufficient to establish the nature of molecular orientation within the monolayer, but the value observed is within comfortable reach of several packing schemes that can be readily envisaged.

In order to make progress, one needs independent information about the tilt angle. This has been obtained in two ways: first, from the change in layer thickness as one adds 3.34 Å to the rod length by inserting one additional bicyclo[1.1.1]pentane cage, and second, from relative infrared intensities.

Results for the metal ion-to-metal ion average thickness of a pair of layers were obtained on multilayer assemblies deposited on an Si wafer. The measurements were performed independently by ellipsometry and by low-angle X-ray diffraction.[29] From these, the average tilt angles were estimated to be 17 ± 5°. The X-ray results suggested that the films are highly ordered. Low-resolution maps of electron density projected into the surface normal clearly showed the location of the metal ions, the region of low electron density in the region of the H···H van der Waals contact in the center of the unit cell, and an intermediate region of oscillating electron density due to the bicyclo[1.1.1]pentane cages (Figure 5-6). In the monolayers of the Cd salt of [4]staffane-3-carboxylic acid, the distance between the unsubstituted terminal bicyclo[1.1.1]pentane cages was only 4.5 Å, and the rod ends must be interdigitated. The Ca salts showed a different packing in that relatively large terminal H···H distances of 6.8 and 7.5 Å were observed in the [3]- and the [4]staffane-3-carboxylate, respectively, and it was concluded that some water probably was occluded in the plane between the unsubstituted rod ends. In contrast, the carboxylate-carrying terminal cages lying across the ionic contact plane were found to lie quite close to each other (6.4 and 5.5 Å,

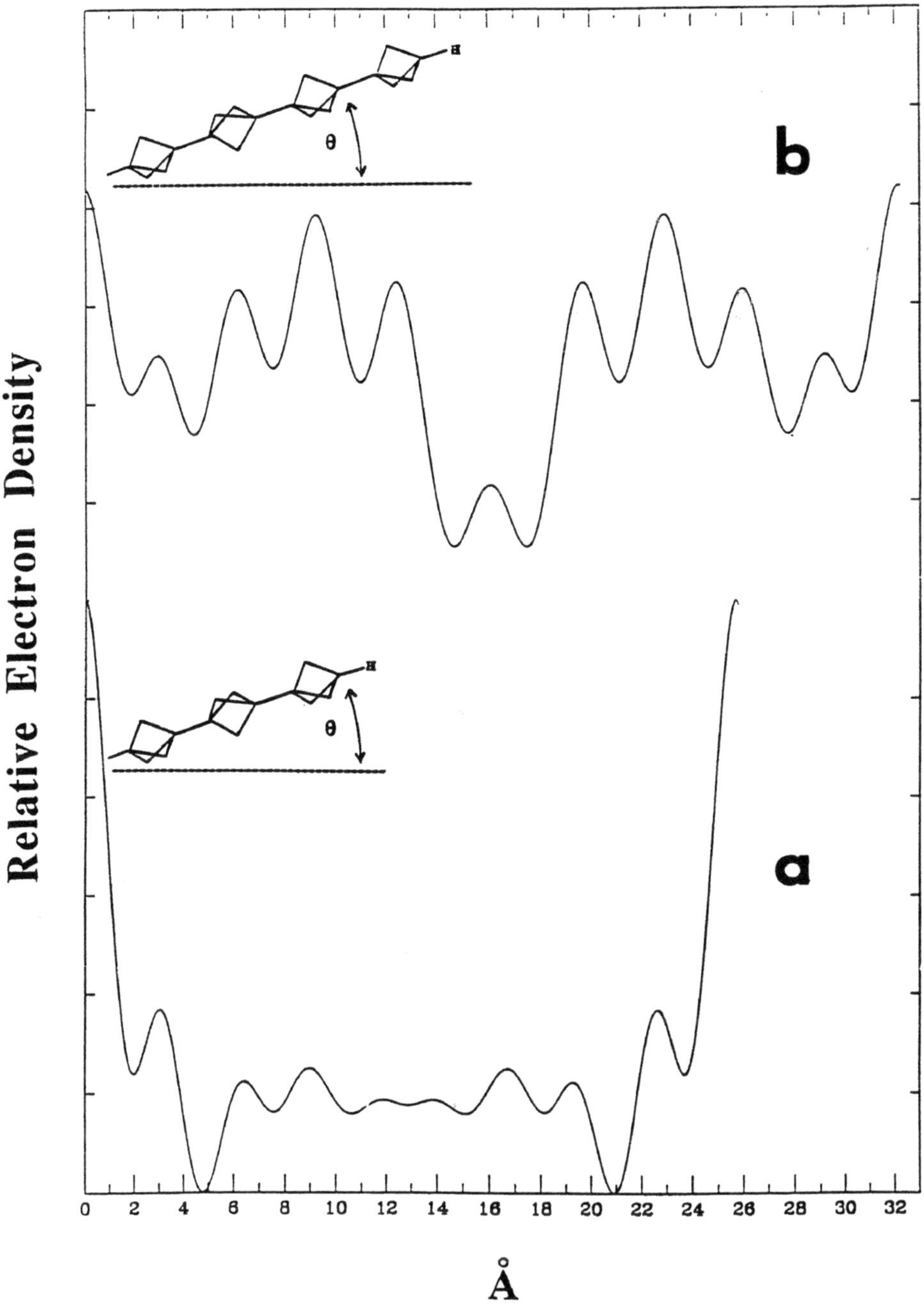

FIGURE 5-6 Projection of the relative unit cell electron densities in Langmuir-Blodgett films of Ca [*n*]staffane-3-carboxylates, (a) $n = 3$ and (b) $n = 4$, onto the surface normal. The vertical scale is arbitrary. (Reproduced by permission from Ref. 29)

respectively). It appears that the cations and anions are meshed and located nearly in the same plane, as proposed for Mn stearate.[30] The difference in the two salts was reflected in shorter metal ion-to-metal ion separation in the Cd salt. Assuming a 17° tilt angle, the effective lengths of the back-to-back arranged rod pairs were 27.0, 33.6, and 32.8 Å for Ca [3]staffanecarboxylate, Ca [4]staffanecarboxylate, and Cd [4]staffanecarboxylate, respectively.

IR spectroscopy has been an independent source of information on the tilt angle of the [*n*]staffane rods in the thin films. The key observable is the ratio of the absorbance due to long-axis polarized transitions to that due to transversely polarized transitions attributed to the vibrations of the [*n*]staffane skeleton. As noted above, the most useful long-axis polarized vibrations are a very strong CH_2 wag located at about 1215 cm^{-1} and a weak CH stretch of the terminal bond at about 2980 cm^{-1}. The intense Fermi doublet at 2880 and 2915 cm^{-1} associated with the CH_2 stretching motion is the most useful of the vibrations that are polarized perpendicular to the long axis.

Vibrations of the carboxylate group were observed as well. However, their relative intensities indicated that the presence of nearby counterions lowers the symmetry of the carboxylate group enough that the directions of the transition moments of the nominally symmetric and antisymmetric CO stretching vibrations no longer coincide with the symmetry axes of the isolated carboxylate group and thus are useless for the present purpose.

Quantitative evaluation is based on the relative intensities of the parallel and perpendicular polarized IR transitions measured on isotropic powder and on the films. In the powder measurement, all transitions are weighted equally. In a measurement on a film taken with the light beam incident normal to a transparent flat substrate (germanium), the observed band intensities are weighted by the squares of the projections of the unit vectors in the transition moment direction into the surface. At grazing angle incidence on an Au-plated surface, they are weighted by the squares of the projections into the surface normal.

Reasonable care needs to be exercised in the analysis to allow for the fact that the transverse vibrations of the [*n*]staffane skeleton are doubly degenerate. As the rod tilts away from the surface normal by angle α, one of the pair of transverse transition moments associated with the two components of the vibrationally excited state remains parallel to the surface, while the other produces a projection into the normal that goes as $\sin \alpha$ and a projection into the surface plane that goes as $\cos \alpha$. A transition moment polarized along the long axis of the rod produces a projection into the surface normal that is proportional to cos α and a projection into the surface that is proportional to sin α.

The tilt angles derived from IR measurements on multilayers of the Cd salt of [4]staffane-3-carboxylic acid are about 20° on the Au substrate and about 30° on the Ge substrate. These results agree very well with those obtained in the ellipsometric and X-ray reflection work on the same multilayer assemblies.

The sensitivity of the IR method is sufficient to perform measurements on a monolayer (Figure 5-7). The results for a monolayer on a Ge surface were close to those obtained for the multilayers, and we believe that the somewhat larger tilt angle reflects a greater roughness of the substrate used rather than any specific property of germanium. On the Au substrate, however, the first layer exhibited a tilt angle that was within experimental error of zero. As shown in Figure 5-7, in this spectrum the perpendicular polarized CH_2 stretches are absent, while the parallel polarized CH_2 wag is strong, as is the terminal CH stretch. The intensity of the latter, due to a single C-H bond, contrasts eloquently with the absence of any intensity in the stretching modes due to the 24 C-H bonds in the CH_2 groups.

The strictly perpendicular alignment of the first layer of [4]staffane-3-carboxylate rods on the metal surface promises to be useful for future attempts to use [*n*]staffane-based monolayers as a base for the construction of more complex structures. For this purpose, terminally functionalized rods are needed. In these, the terminal bonds would permit the anchoring of further groups in a direction well defined to be normal to the surface. As noted above, we have synthesized [*n*]staffane-3-carboxylic acids with a terminal *p*-iodophenyl group, and hope that these will yield monolayers in which the C-I bonds are arranged perpendicular to the surface. We have already performed the first experiments with terminally iodinated rods based on oligomeric carboranes, which produce Langmuir-Blodgett films with a substantially larger area per molecule, about 40 Å^2 per molecule,[6] and are presently attempting to dissociate the C-I bonds by UV irradiation. Ultimately, the availability of a series of rods of various diameters and lengths should offer considerable freedom in the choice of monolayers for the anchoring of more complex structures.

Self-Assembled Monolayers

While our goal of producing [*n*]staffane-based Langmuir-Blodgett films rigidly and usefully functionalized on their outer surface has not yet been reached, a parallel effort based on the self-assembly of thiols on

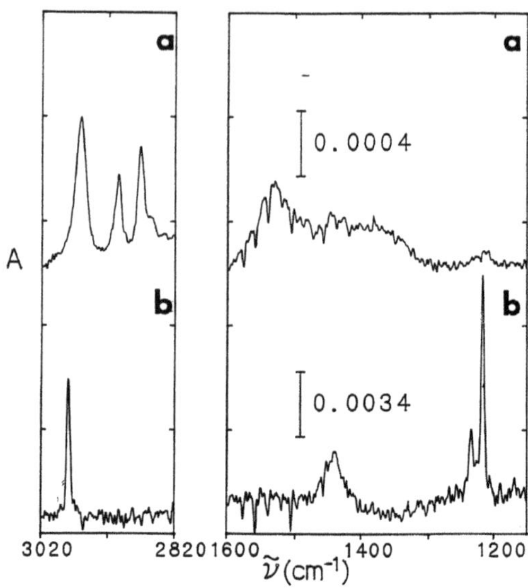

FIGURE 5-7 IR spectra of a Cd [4]staffane-3-carboxylate monolayer: (a) normal incidence on Ge; (b) grazing incidence on Au-covered glass. (Reproduced by permission from Ref. 29)

an Au surface has already produced such a monolayer.[31] Here the target was a surface functionalized with the sulfhydryl (SH) groups. These should be relatively reactive and useful for the attachment of a variety of substituents.

The formation of self-assembled monolayers of alkanethiols by adsorption from solution onto the surface of metals such as Cu[32] and Au[33] is facile and has been extensively studied. Our initial attempts to produce well-ordered monolayers by the adsorption of [n]staffane-3,3$^{(n-1)}$-dithiols on an Au surface gave disappointing results. The layers that formed did not block the Au electrode well, and their IR spectra, analyzed as described above for Langmuir-Blodgett films, showed that the average inclination of the rod axis was about halfway between normal and parallel to the surface. Contact angles for water were very high, about 80°. It appeared likely that only about half of the molecules adsorbed in the desired fashion, through one S atom and normal to the surface, while the other half adsorbed through both S atoms, with the [n]staffane rods parallel to the surface. This could be expected to lead to a poorly ordered, poorly blocking, and quite hydrophobic surface layer.

The obvious remedy was to deactivate one of the SH groups on each rod. An easy way to do this was suggested by the initial mode of synthesis of these [n]staffane derivatives[2] and is summarized in Figure 5-8. As described above, the synthesis produces the thiol acetates **1**(n = 1 − 5, X = Y = $SCOCH_3$) rather than the free thiols. Thus, all that was needed was to perform only partial hydrolysis and to isolate the singly protected thiol-thiol acetates **1**(n = 1 − 5, X = SH, Y = $SCOCH_3$).[31] With the exception of the first member of the series, n = 1, these indeed readily form stable and perfectly blocking self-assembled monolayers on the surface of an Au electrode. Most of the experiments were performed with rods containing two or three bicyclo[1.1.1]pentane cages, n = 2 or 3.

Grazing-incidence IR measurements showed a strong CH_2 wagging peak near 1215 cm^{-1} and negligible intensity for the CH_2 stretching-scissoring Fermi doublet at 2880 and 2915 cm^{-1} (Figure 5-9). This implies that the [n]staffane rods are oriented normal to the surface within the experimental error, as hoped for. It appears most likely that the S atom is centered above a triangle of Au atoms in a [111] surface and thus has a tetrahedral environment.

The C-H stretching IR peaks due to the methyl group are difficult to separate from those of the [n]staffane rod. However, the orientation of the acetyl groups located on the outer surface of the monolayer can still be determined from the IR intensity of the strong carbonyl stretching vibration located near 1700 cm^{-1}. A priori, one would expect it to be either cis planar, with the CSCO dihedral angle ϕ equal to zero (the O atom pointing toward the Au surface and the CH_3 group outward), or trans planar, with ϕ equal to 180° (O outward and CH_3 toward the Au surface).

The IR intensity of the carbonyl stretch relative to the CH_2 wag, which is polarized normal to the surface, is reduced by a factor of about one-half with respect to its value in an isotropic powder. This factor is equal to $<\cos^2 \sigma>$, where σ is the angle between the C==O bond direction and the surface normal, and the pointed brackets indicate ensemble averaging. Assuming that all of the acetyl groups in the surface have similar conformations, trigonometry then shows that the IR intensities are compatible with two possible orientations. The dihedral angle ϕ is either close to zero (cis) or close to 130° (gauche). The trans conformation (ϕ = 180°) is excluded. The gauche conformation, although compatible with the IR intensities, would be quite different from the conformations known from single-crystal X-ray structures of [n]staffane thiol acetate derivatives (Figure 5-2),[20] would have strongly reduced conjugation between the S and carbonyl groups, and appears improbable. The cis conformation is most likely by far, and we conclude that the methyl groups point outward, as they do in the single crystals of $3,3^{(n-1)}$-bisacetylthio[n]staffanes[20] This is nicely compatible with the relatively

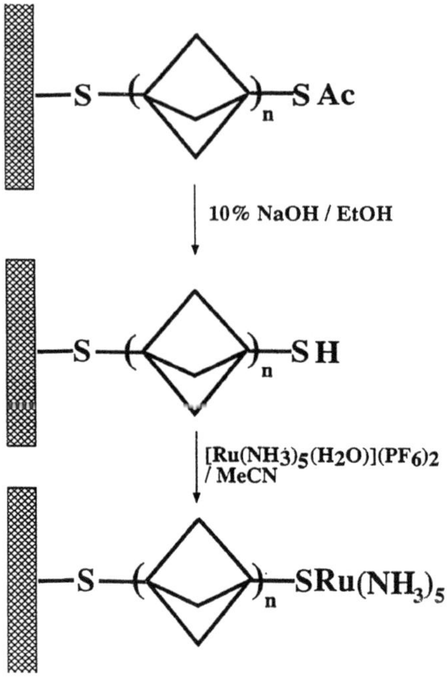

FIGURE 5-8 Schematic representation of the chemical transformations of a self-assembled monolayer. (Reproduced by permission from Ref. 31)

high contact angles for water on this surface, about 75°. This value can be compared with the value of 70° reported[34] for a layer self-assembled from the aliphatic thiol-thiol acetate, $HS(CH_2)_{12}SCOCH_3$.

Next, the acetyl group was hydrolyzed from the film surface by brief exposure to a dilute ethanolic solution of NaOH. The only detectable change that this caused in the IR spectra was the expected disappearance of the carbonyl stretching peak at 1700 cm^{-1} (Figure 5-9). In particular, the perpendicular polarized CH_2 stretches remain undetectable. The electrode-blocking properties of the film remain unchanged as well. We conclude that removal of the protecting groups does not significantly perturb the structure of the monolayer. The [*n*]staffane rods remain oriented normal to the surface, which is now decorated by SH groups, as originally desired. Their presence is reflected in the much reduced water contact angle, about 40°.

The sulfhydrylated surface is now ready for further functionalization. The first types of experiments we wished to perform were electrochemical, and we decided to start with the ruthenium pentaammine substituent as a model entity susceptible to oxidation and reduction. This group

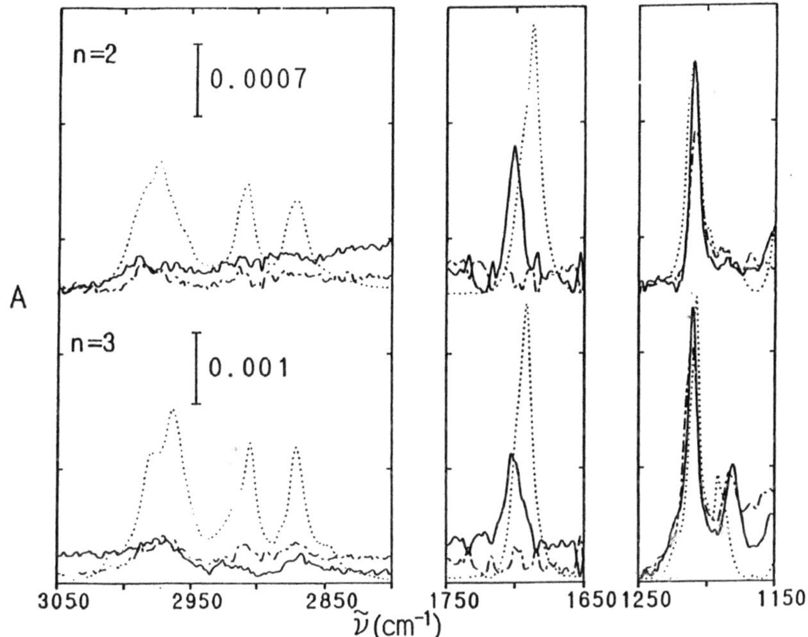

FIGURE 5-9 Grazing incidence IR absorption spectra of a monolayer of the thiol-thiol acetate **1** (X = SH, Y = $SCOCH_3$) as initially self-assembled (solid line) and after hydrolytic removal of the acetate group (dash-dotted line) on an Au substrate. Dotted line: isotropic spectrum of **1** (X = SH, Y = $SCOCH_3$) in a KBr pellet, plotted on a scale such as to make the intensity of the 1210 cm^{-1} peak equal to that in the self-assembled film. (Reproduced by permission from Ref. 31)

was introduced by an overnight immersion of the sulfhydrylated surface in a dilute solution of $[Ru(NH_3)_5(H_2O)](PF_6)_2$ in acetonitrile. The IR spectrum of the resulting monolayer (Figure 5-10) is dominated by features associated with the ammonia ligands on the ruthenium atoms: the broad NH stretching band between 3000 and 3400 cm^{-1}, the degenerate deformation band near 1600 cm^{-1}, and the symmetric deformation band at 1333 cm^{-1}. The CH_2 stretching bands are still undetectably weak and the CH_2 wag is easily detectable, proving that the [*n*]staffane rods are standing upright on the surface.

Very similar monolayers were obtained by self-assembly of [*n*]staffane-3,$3^{(n-1)}$-dithiols carrying a ruthenium pentaammine group on one of the thiols, **1**(X = SH, Y = $SRu(NH_3)_5$). Only the IR intensity of the CH_2 wag, characteristic of the [*n*]staffane moiety, was weaker by a factor of at least 2 (Figure 5-10). This can be understood on consideration of the relative size of the ruthenium pentaammine group, with a radius of

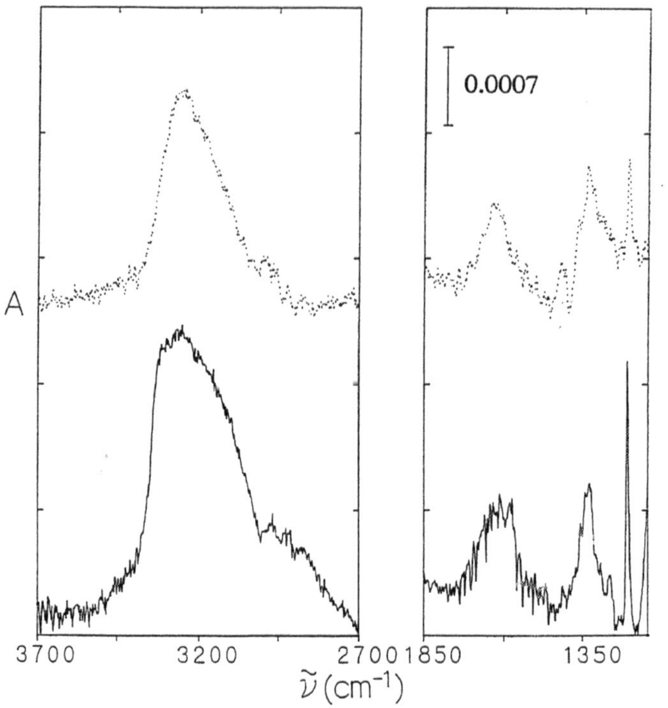

FIGURE 5-10 Grazing incidence IR absorption spectra of ruthenium pentaammine carrying self-assembled monolayers on an Au substrate. Solid line: film obtained by rutheniation of a sulfhydrylated surface. Dotted line: film assembled from preruthenated thiol, **1** [X = SH, Y = $SRu(NH_3)_5$]. (Reproduced by permission from Ref. 31)

about 3.5 Å, and of the staffane rod, with a diameter close to 2.9 Å. On rutheniation of the sulfhydrylated surface, at least half of the SH groups must be left unfunctionalized in order to accommodate the ruthenium pentaammine groups properly. In the self-assembly of [*n*]staffane rods that already carry the ruthenium pentaammine group on one end, the same spatial problem arises. Since every rod carries a ruthenium pentaammine group, they cannot pack as densely as they did in all of the films discussed so far, and the observed intensity of the CH_2 wag must be lower, as observed. It is not obvious how the voids in the film are filled; the counterions are an obvious possibility. They are apparently filled well, since the blocking properties of the films produced by self-assembly of the singly rutheniated rods are excellent and are indistinguishable ffom those of films that were produced by rutheniation of the sulfhydrylated surfaces. There is no evidence of water in the spectra of dried films. The monolayers were quite hydrophilic and had water contact angles of about 40-60°.

The blocking properties of the films were evaluated by three procedures. One of these was cyclic voltammetry on a solution containing ferrocyanide anions (Figure 5-11), the second was chronoamperometry on a solution containing ruthenium hexaammine cations, and the third was a measurement of a decrease in the gold oxide reduction wave by the blocking layer.

The electrochemical behavior of monolayers whose outer surface carried ruthenium pentaammine groups was the same, regardless of how they were prepared: either by functionalization of a sulfhydrylated surface or by self-assembly of rods carrying a preattached ruthenium pentaammine substituent on one end.

On a glassy carbon electrode, which does not adsorb thiols, acidic solutions of the singly rutheniated rod **1**(n = 2, X = SH, Y = $SRu(NH_3)_5$) showed reversible oxidation-reduction behavior. Immobilized on the surface of an Au electrode, the rutheniated thiol yielded peak currents for oxidation and reduction that were linear with scan rate, and the peak potential was 0.51 V vs. SCE (v = 200 mV/s), about 60 mV more positive than the solution species (Figure 5-12).

The difference in oxidation-reduction potentials suggests that Ru(II) in the film is stabilized with respect to Ru(III), which may be attributed to increased electrostatic repulsions between the Ru centers in the +3 oxidation state. Alternatively, the difference in the potentials could be due to differences in specific ion pairs formed with the perchlorate anion from the supporting electrolyte[35] After background correction, the charge passed under the anodic wave was measured to be 1.2×10^{-10} mol/cm^2. The roughness (increase in surface area) of the polycrystalline Au substrate was 1.6 ± 0.2, as determined from the charge passed for the reduction of an adsorbed iodine monolayer on bare Au. The adsorbed coverage corresponded to about 40% of that expected for a monolayer with the densest commensurate packing of the [n]staffane rods on gold[111] [83 Å^2 per $SRu(NH_3)_5$]. This relatively low coverage is probably due to electrostatic repulsions among the positively charged Ru centers and to the need to accommodate counterions. If all of the divalent counterions or half of the monovalent ones are part of the surface monolayer, the observed coverage corresponds to complete surface coverage within experimental error.

Measurements of the rate of the oxidation and reduction processes at the immobilized layer by the potential jump method gave results that were not compatible with electron transfer across the [n]staffane rod being the rate determining step. The kinetic behavior was nonexponential.

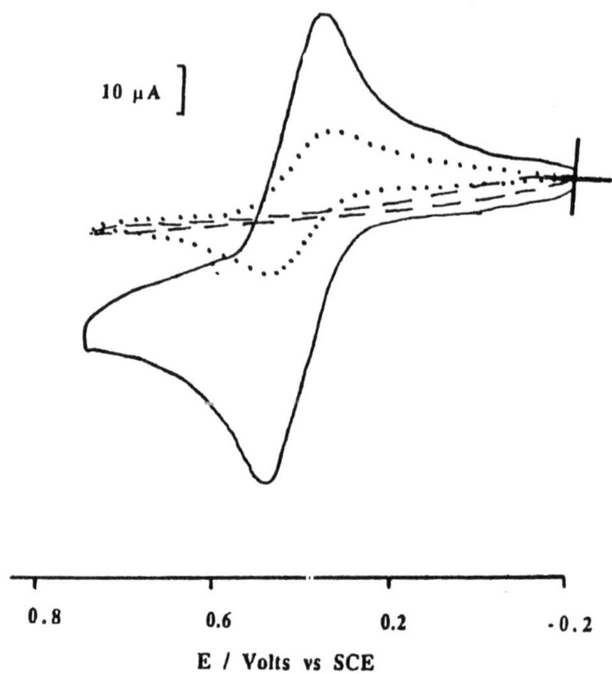

FIGURE 5-11 Cyclic voltammogram of a 1.2 mM solution of $K_4Fe(CN)_6$ at a bare Au electrode (solid line), an Au electrode coated with a self-assembled monolayer of the thiol-thiol acetate **1**(n = 4, X = SH, Y = $SCOCH_3$) (dashed line), and an Au electrode coated with a self-assembled monolayer of the dithiol **1**(n = 4, X = Y = SH) (dotted line). Scan rate 50 mVs^{-1}, SCE reference electrode. (Reproduced by permission from Ref. 31)

The rates were unreasonably high and decreased only slowly as the rods became longer (at the half-wave potential, very roughly 1600, 900, and 500 s^{-1} for the reduction of the Ru center on the outer surface of monolayers prepared from [2]-, [3]-, and [4]staffanedithiols). It appears likely that the electrons reach the outer surface at defect sites, at a rate that depends only weakly on monolayer thickness, and subsequently jump from one Ru center to the next along the outer surface. If this explanation is correct, a decrease in the density of functionalization of the surface with the $Ru(NH_3)_5$ units, which would slow down electron transfer parallel to the surface or eliminate it altogether, should have a large effect on the observed kinetics. It is to be hoped that in the limit of low coverage, the transfer of electrons across the length of the [n]staffane rod will become the rate-determining step. Measurements of this type are underway at present.

If meaningful heterogeneous electron transfer rates across a blocked

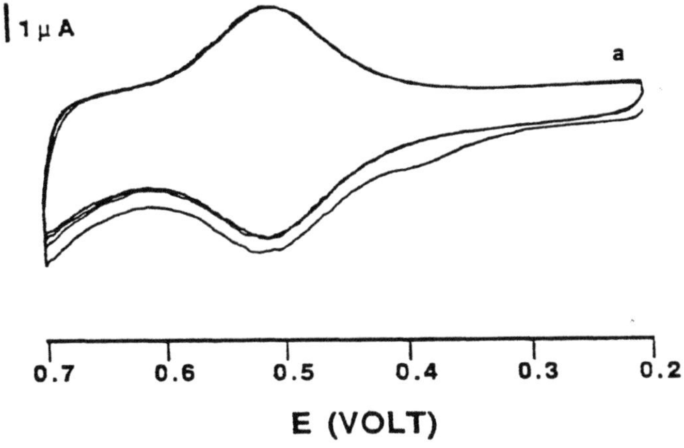

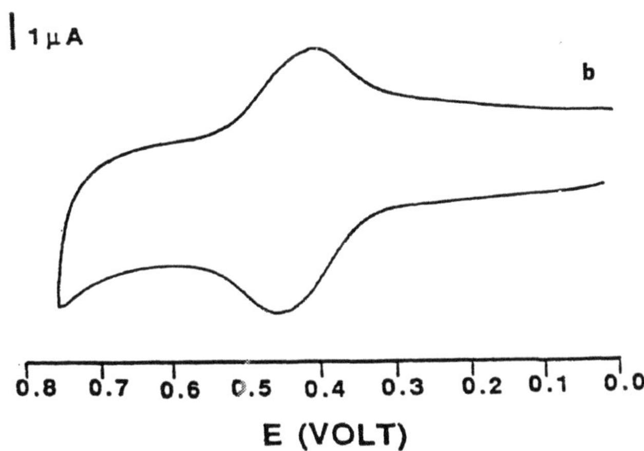

FIGURE 5-12 Cyclic voltammogram of (a) a self-assembled monolayer of **1**[n = 2, X = SH, Y = $SRu(NH_3)_5$] on a polycrystalline Au substrate of 0.18 cm^2 geometric area and (b) a 2 mM solution of **1**[n = 2, X = SH, Y = $SRu(NH_3)_5$] with a glassy carbon electrode in 1 M $HClO_4$, v = 200 mVs^{-1}, SCE reference electrode. (Reproduced by permission from Ref. 31)

electrode can be measured, it will be possible to compare them with the homogeneous electron transfer rates mentioned above. These were estimated for the same [n]staffane-3,$3^{(n-1)}$dithiols carrying a ruthenium pentaammine group on each S from the properties of the charge-transfer band observed in the mixed-valence Ru(II)-Ru(III) species. It will be very interesting to see whether the electron transfer rates are high enough for imaging the monolayers by scanning tunneling microscopy.

CONCLUSIONS AND OUTLOOK

Both Langmuir-Blodgett and self-assembled monolayers can be built from suitably functionalized [*n*]staffanes, and the long axes of the rods are perpendicular to the surface. Monolayers of both types have been provided with functionalities at the outer surface, and it is believed that the rigidity of the rods will be helpful in preventing undesirable inter-substituent interactions. Rod packing in multilayers is different and approximates that in single crystals of [*n*]staffanes. Much work remains to be done before the arrangement of the rods in the plane of the surface is fully understood.

Investigations of further manipulation of the functionalities on the outer surface of the monolayers are underway. Thus, removal of iodine atoms by UV irradiation should produce an outer surface adorned with free radical centers, with obvious further uses. The attachment of suitable molecular-recognition groups at the sulfhydrylated surface may produce an electrode surface that is nearly perfectly blocking for species present in solution, and yet permits reduction of species that are recognized and selectively bound from the solution to the functionalized monolayer on the surface of the electrode. Such a functionalized electrode would be of interest both for selective sensing and for selective preparative electrochemistry.

However, as pointed out above, we view even the most intriguing objects of this chapter, supramolecular rod assemblies, only as an interesting intermediate stage on the way to our ultimate goal: fully covalently bonded thin-layer assemblies carrying active groups at preselected locations.

Acknowledgment

I am grateful to a large number of very gifted collaborators who obtained the results discussed in this chapter. During the stay of my research group at the University of Texas in Austin, I benefitted greatly from a collaboration with Prof. A.J. Bard and several members of his research group. The names of all of those who contributed are listed in the references. Our work on has been supported by the National Science Foundation: grants CHE-8901450, CHE-9020896, DMR-8807701, DMR-5114070, CHE-9022151, and CHE-9121643. Additional support was provided by the donors of the Petroleum Research Fund of the American Chemical Society (24932 AC). During my stay at the University of Texas, further support was provided by the Robert A. Welch Foundation (Grant 1068) and by the Texas Advanced Technology Program.

REFERENCES

1. J. Michl, P. Kaszynski, A.C. Friedli, G.S. Murthy, H.-C. Yang, R.E. Robinson, N.D. McMurdie, and T. Kim, in *Strain and Its Implications in Organic Chemistry*, A. de Meijere and S. Blechert, Eds., *NATO ASI Series*, Vol. 273 (Kluwer, Dordrecht, the Netherlands, 1989), p. 463.
2. P. Kaszynski, A.C. Friedli, and J. Michl, *J. Am. Chem. Soc.*, **114**, 601 (1991).
3. Z. Wu, S. Lee, and J.S. Moore, *J. Am. Chem. Soc.*, **114**, 8730 (1992).
4. P. Kaszynski and J. Michl, *J. Am. Chem. Soc.*, **110**, 5225 (1988).
5. K. Hassenrück, G.S. Murthy, V.M. Lynch, and J. Michl, *J. Org Chem.*, **55**, 1013 (1990).
6. J. Müller, K. Baše, T.F. Magnera, and J. Michl, *J. Am. Chem. Soc.*, **114**, 9721 (1992).
7. X. Yang, W. Jiang, C.B. Knobler, and M.F. Hawthorne, *J. Am. Chem. Soc.*, **114**, 9719 (1992).
8. H.E. Zimmerman, R.K. King, and M.B. Meinhardt, *J. Org. Chem.*, **57**, 5484 (1992).
9. P.E. Eaton and J. Tsanaktsidis, *J. Am. Chem. Soc.*, **112**, 876 (1990).
10. P.W. Kenny and L.L. Miller, *J. Chem. Soc., Chem. Commun.*, 84 (1988).
11. B.H. Robinson and N. Seeman, *Protein Eng.*, **1**, 295 (1987) and references therein.
12. K.B. Wiberg and F.H. Walker, *J. Am. Chem. Soc.*, **104**, 5239 (1982).
13. K. Semmler, G. Szeimies, and J. Belzner, *J. Am. Chem. Soc.*, **107**, 6410 (1985).
14. P. Kaszynski and J. Michl, *J. Org. Chem.*, **53**, 4593 (1988).
15. T. Janecki, S. Shi, P. Kaszynski, and J. Michl, *Collect. Czech. Chem. Commun.*, **58**, 89 (1993).
16. A. Koch and J. Michl, unpublished results.
17. F.N. Tebbe, P.M. Garrett, and M.F. Hawthorne, *J. Am. Chem. Soc.*, **90**, 869 (1968).
18. S. Papetti, C. Obenlaud, and T.L. Heying, I&EC Products Res. & Dvpt., **5**, 334 (1966).
19. L.I. Zakharkin and A.I. Kovredov, *Zh. Obshch. Khim.*, **44**, 1860 (1974); *Izv. Akad. Nauk SSSR, Ser. Khim*, 1428 (1973).
20. A.C. Friedli, V.M. Lynch, P. Kaszynski, and J. Michl, *Acta Cryst.*, **46**, 377 (*B* 1990).
21. G.S. Murthy, K. Hassenrück, V.M. Lynch, and J. Michl, *J. Am. Chem. Soc.*, **111**, 7262 (1989).
22. H.A. Bent, *Chem. Rev.*, **61**, 275 (1961).
23. M.S. Gudipati, S.J. Hamrock, V. Balaji, and J. Michl, *J. Phys, Chem.*, **96**, 10165 (1992).

24. E.W. Thulstrup and J. Michl, *Spectrochim. Acta,* **44A,** 767 (1988).
25. R. Gleiter, K.-H. Pfeifer, G. Szeimies, and U. Bunz, *Angew. Chem. Int. Ed. Engl.,* **29,** 413 (1990).
26. A.J. McKinley, P.N. Ibrahim, V. Balaji, and J. Michl, *J. Am. Chem. Soc.,* **114,** 10631 (1992).
27. J. Michl, A.C. Friedli, Y.S. Obeng, and A.J. Bard, *Proceedings of the Sixth International Conference on Energy and Electron Transfer,* J. Fiala and J. Pokorný, Eds., (Charles University, Prague, 1989). The rates reported in the proceedings are too high due to the use of an erroneous literature equation for data treatment. The correct rate constants are 1×10^7 and 2×10^6 s^{-1} for [1]3 and [2]3, respectively, in 0.1 M DCl in D_2O.
28. C.A. Stein, N.A. Lewis, and G. Seitz, *J. Am. Chem. Soc.,* **104,** 2596 (1982).
29. H.C. Yang, T.F. Magnera, C.-M. Lee, A.J. Bard, and J. Michl, *Langmuir,* **8,** 2740 (1992).
30. M. Pomerantz and A. Segmüller, *Thin Solid Films,* **68,** 33 (1980).
31. Y.S. Obeng, M.E. Laing, A.C. Friedli, H.C. Yang, D. Wang, E.W. Thulstrup, A.J. Bard, and J. Michl, *J. Am. Chem. Soc.,* **114,** 9943 (1992).
32. L.C.F. Blackman and M.J.S. Dewar, *J. Chem. Soc.,* 171 (1957).
33. R.G. Nuzzo and D.L. Allara, *J. Am. Chem. Soc.,* **105,** 4481 (1983).
34. G.M. Whitesides and P.E. Laibnis, *Langmuir,* **6,** 87 (1990).
35. J. Redepenning, H.M. Tunison, and H.O. Finklea, *Langmuir,* **9,** 1404 (1993).

6
Graphite: Flat, Fibrous and Spherical

JOHN A. JASZCZAK

INTRODUCTION

The paradigm for the structure of graphite is that of a staggered stacking of flat layers of carbon atoms (Figure 6-1). Individual layers, sometimes referred to as *graphene* sheets,[1] are weakly bonded to each other and are composed of strongly bonded carbon atoms at the vertices of a network of regular hexagons in a honeycomb pattern.[2] Both the properties and the morphology of graphite reflect its highly anisotropic structure. Due to the strong bonding within layers and the weak bonding between layers, the growth of graphite takes place predominantly along the edges of the layers (perpendicular to the *c* axis) and only very slowly normal to the layers (parallel to the *c* axis). As a result of the growth rate anisotropy, the anisotropic surface energy, and the crystallographic symmetry, the expected morphology for graphite crystals is that of tabular hexagonal prisms.[3] However, well-formed natural crystals, such as shown in Figure 6-2, are rare,[4] and it has been said that near-ideal crystals of graphite may be rarer than diamonds.[5] Well-formed, laboratory-grown crystals of graphite are also uncommon. During the 1960s, graphite crystals from Ticonderoga, New York, and Sterling Hill, New Jersey, became a standard of perfection for experiments and for comparison with laboratory-grown crystals.[6,7]

One of the reasons for the rarity of such graphite crystals is the extreme softness and flexibility of graphite due to the ease with which the carbon sheets in its structure can glide past one another under stress. In

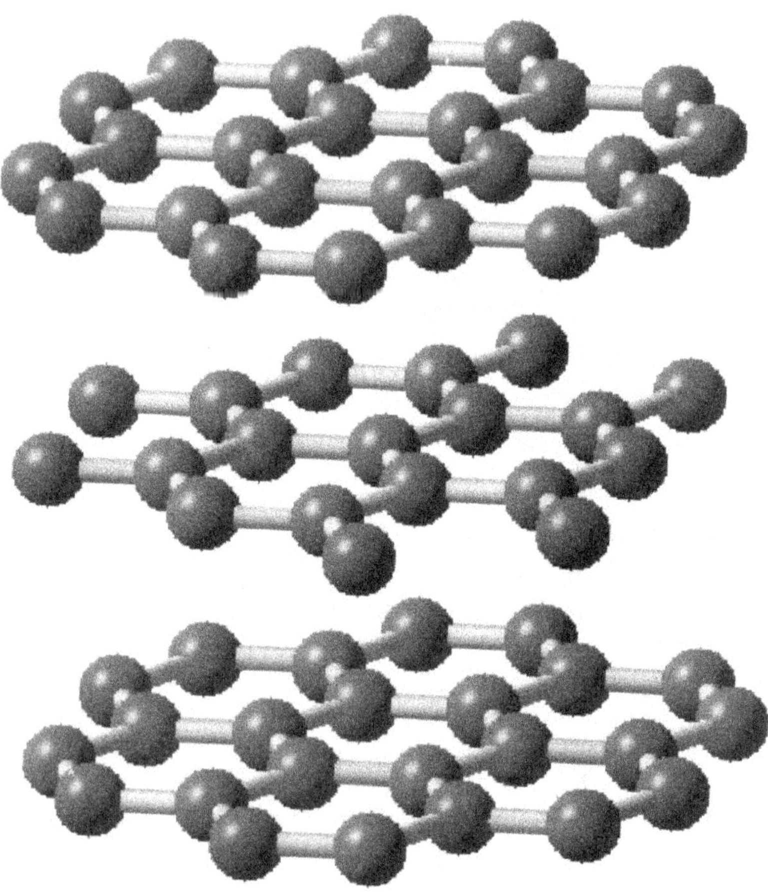

FIGURE 6-1 A portion of the graphite structure showing three planes of atoms in the hexagonal ... ABAB ... stacking sequence with space-group symmetry $P6_3/mmc$. The C-C bond distance in the planes is 1.42 Å, while the interplanar spacing is 3.40 Å. (Image created using MacMolecule; ©University of Arizona)

FIGURE 6-2 Well-formed graphite crystal (1 mm) showing tabular hexagonal morphology in calcite ($CaCO_3$) from Crestmore, Riverside County, California. JAJ specimen No. 888.

fact, graphite is known to occur, both naturally and synthetically, in a variety of curved morphologies, including cylindrical and spherical, at length scales ranging from near atomic to macroscopic.[8] Perhaps another reason for the rarity of well-formed graphite crystals and for the existence of curved structures is related to the nucleation and growth processes of graphite. Since flat but finite sheets of honeycomb-bonded carbon atoms must have broken bonds at their edges, flat graphene sheets may not be the most stable form of carbon in very small systems.[9,10] On a small enough scale, such as during the early stages of nucleation, the thin, flexible graphene sheets could easily become curved in order to minimize their surface energy or to satisfy additional bonds at these edges.[11] Curving of the graphitic sheets at the smallest scale can be accomplished by the addition of one or more pentagons in the carbon sheet of hexagons. The addition of 12 pentagons in a sheet allows the sheet to close on itself and to form fullerenes and other spherical carbon clusters.[12,13] The smallest closed carbon cage with no two pentagons touching each other is the fullerene C_{60}. The flexibility of graphitic sheets and the inclusion of pentagons therein have been proposed to explain the growth of graphitic carbon fibers with a cone-helix structure[14-17] (see below).

Inspired by the existence of fullerenes[18] and larger curved graphites, we here briefly review the literature on graphites occurring as fibers and spheres, from microscopic to macroscopic. Also summarized for the first time are the geological occurrences of spherical graphite.

FILAMENTOUS GRAPHITE

Due to the unique strength of the in-plane carbon-carbon bond in graphite, carbon fibers are used to produce some of the strongest materials for their weight known. According to Hillert and Lange,[19] the filamentary growth of graphite was observed over a century ago. Because of their technological importance, graphite fibers have been the subject of a great deal of research since the 1940s.[19,20] Recent reviews of filamentous carbon have been given by Baker and Harris,[21] Dresselhaus et al.,[1] Tibbetts,[22] and Rodriguez.[23] Commercially produced carbon fibers (known as *ex-polymer* fibers) are made by heat treatment (graphitization) of extruded polymer precursors. These fibers have a variety of graphitic microstructures, depending on processing conditions and the polymer precursor, but are typically rather disordered, especially in cross section. More highly ordered filaments are formed from catalytic chemical vapor deposition (CCVD). Such filaments are usually formed as a result of the decomposition of carbon-rich gases in the presence of some solid catalyst at temperatures usually in the range of 350 °C to 2500 °C. The filaments typically range in size from 0.01 to 200 μm in diameter, although the thicker fibers can reach up to 5 cm in length. The filaments can be straight, curved, bent, or twisted. Using a two-stage vapor-growth process of first nucleating and lengthening thin fibers with nanometer-sized, iron-rich catalyst particles and then thickening them by increasing the carbon potential, straight fibers have been grown up to 30 cm long.[22] CCVD filaments are thought to be composed of concentric rings of graphene sheets with the c axis (sheet normal) normal to the fiber axis (Figure 6-3a). After heat treatment above 3000 °C, the filaments can develop facets (Figure 6-3b).[1,24] The most ordered carbon fibers (termed *graphite whiskers*) are grown in a carbon arc under a high pressure of inert gas (Ar or He). These fibers typically consist of graphene sheets rolled up in a scroll-like fashion, with the c axis roughly perpendicular to the fiber axis (Figure 6-3c).[20] Filamentous carbon occurs in a variety of structures, however, in which the c axis can be perpendicular to, parallel to, or inclined with respect to the fiber axis (Figure 6-3). Due to their highly ordered structures, arc-grown graphite whiskers are far superior in strength to com-

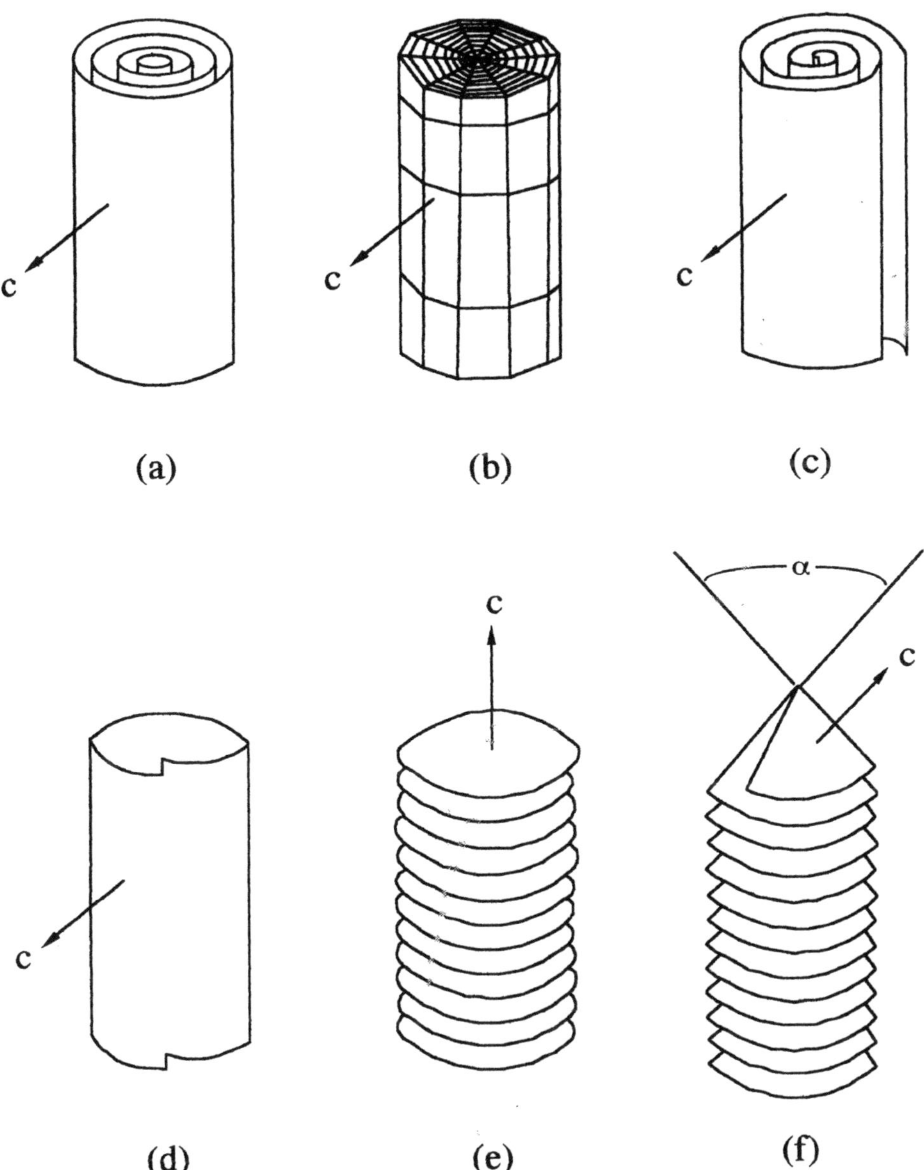

FIGURE 6-3 Schematic representation of various fibrous graphite morphologies. The concentric tube, or tree-ring, types (a) and (b), are commonly produced by CCVD, as well as by carbon arcs. Part (b) shows the effect of high-temperature heat treatment of CCVD fibers (a) resulting in faceting. Scroll types (c) can be produced in carbon arcs. Types (a) through (d) have the *c* axis (graphene plane normal) normal to the fiber axis. These types often have helicity around the fiber axis leading to a growth ledge at the fiber ends (d). The *c* axis in (e) is parallel to the fiber axis and may or may not contain a screw dislocation parallel to the fiber axis. Cone-helix whiskers (f) have the *c* axis tilted with respect to the fiber axis.

mercially available carbon fibers and have been of great interest for potential commercial applications.[1]

One of the most recent additions to the list of filamentous graphite types are carbon-arc-generated hollow carbon tubes, sometimes referred to as *buckytubes,* that are only nanometers in diameter.[25,26] These tiny fibers are composed of concentric tubes of graphene (Figure 6-3a) and do not appear to be scroll-like. Furthermore, the hexagons in each layer are arranged in a helical fashion about the filament axis, producing a type of growth step at the fiber end that may expedite growth parallel to the fiber axis (Figure 6-3d). Annealing of such nanotubes in an intense electron beam results in their becoming more spherical by the addition of pentagons to the network of hexagons, and ultimately results in the formation of carbon "onions"—concentric spheres of graphene layers.[27–29] Carbon nanotubes have also been filled with liquid lead by capillary action, giving rise to the possibility of nano-fabrication of wires, as well as new electrical and mechanical properties.[30] While the first buckytubes were composed of 2 to 50 coaxial graphene sheets (not rolled like a scroll), nanotubes of only a *single* atomic layer thick have recently been synthesized from the gas phase in the presence of either Fe[31] or Co.[32]

Naturally occurring graphite whiskers have been reported by Patel and Deshapande.[33] Several whiskers were observed with graphite crystals from Ticonderoga, New York, that were up to 1 mm long and rather thick (65 to 125 μm in diameter). Unlike most filamentous graphite, these naturally occurring whiskers grew with the *c* axis parallel to the fiber axis (Figure 6-3e). Recently, ribbon-like graphite filaments have been grown by disproportionation of carbon monoxide by small, iron-rich particles at the growing ends of the whiskers.[34] These whiskers were shown by high-resolution transmission electron microscopy (HRTEM) to be highly ordered and "novel' in that they also grew with the fiber axis parallel to the *c* axis of the graphite.

Carbon whiskers made of the fullerene C_{60} comprise another class. Krätschmer et al.[35] first produced macroscopic amounts of C_{60} from arc-generated carbon soot. The C_{60} was extracted from the carbon soot in benzene and crystallized as thin platelets, rods, and stars with hexagonal symmetry. The structure of pure C_{60} crystals seems actually to have the C_{60} molecules centered on face-centered cubic (fcc) lattice sites.[36-39] Recently, Yosida has grown whisker crystals of C_{60} in a benzene and hexane solvent.[40] These whiskers, 60 μm in diameter and 300 μm long, have an [001] direction as the fiber axis and show a faceted morphology consistent with an fcc structure.

Another class of carbon filaments are known to have helical structures in which the local normals of the graphene sheets are inclined with respect to the fiber axis.[17,19] Amelinckx et al.[17] have recently characterized the structure of graphite whiskers that break to form conical surfaces and have proposed that individual whiskers grow around a type of defect called a *disclination*.[41] To form a disclination, consider a single graphene sheet of carbon atoms with a cut in the sheet from the edge to the middle to the sheet. By making the cut edges overlap, a cone shape is formed from the sheet, and continued growth on the exposed edge yields a cone-helix structure (Figure 6-3f). The larger the overlap of the cut edges, the smaller the apex angle α. In the limit of no overlap, $\alpha = 180°$, and the disclination becomes a classical screw dislocation, provided that the cut edge is still shifted sufficiently to form a step on the graphene sheet for continued growth. The exposed cut edge serves as a nucleation site for additional growth, as is expected, but now with an overall component of the fiber's growth direction normal to the graphene sheets. Amelinckx et al. further proposed that the disclination core is a channel of stacked pentagons in the otherwise hexagonal graphene sheet. Double and Hellawell[14-16] had previously proposed the same mechanism for the formation of cone-helix structures, and noted that certain overlap angles (and thus certain α angles) form low-energy twist grain boundaries and would be preferred over other overlap angles. In particular, for the often observed value for the apex angle $\alpha = 139.9°$, the graphene planes are rotated about the *c* axis by 21.8° and form low-energy twist boundaries. Double and Hellawell further proposed that numerous cone-helix fibers could nucleate from a common center (perhaps a fullerene or carbon onion) to form spherical graphite (Figure 6-4), a common and important morphology for graphite in cast iron and metallic solutions.

SPHERICAL GRAPHITE

The 22 October 1992 cover of *Nature* showed a stunning micrograph of a carbon onion made by Daniel Ugarte[27,28] that was captioned "The Ultimate Fullerene?". In his article describing the formation of quasi-spherical particles composed of highly ordered, concentric graphitic shells up to several microns in diameter by electron-beam irradiation of carbon soot, Ugarte suggested that planar graphite may not be the most stable allotrope of carbon in systems of limited size.[28] His results presented clear experimental evidence of a spontaneous tendency for graphene or graphite to include pentagons in its network of hexagons and form curved structures. Nested polyhedral graphene cages similar to Ugarte's onions had been observed by Iijima in vacuum-deposited, amorphous carbon films[12] (Figure 6-5). It has been suggested that in the

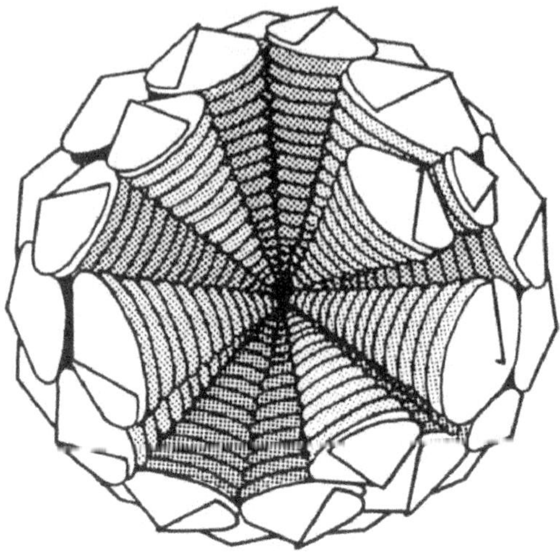

FIGURE 6-4 Sketch of numerous cone-helix graphene sheets radiating from a central nucleus to form a spherical morphology. [From Figure 3(a) of Ref. 15. Used by permission]

FIGURE 6-5 Nearly spherical shells of carbon in amorphous carbon films found by Iijima. The arrows indicate radial regions where the carbon shells show interruptions (see also Figure 3 of Ref. 13). The marker at the bottom right represents 20 Å. [From Figure 5(b) of Ref. 42. Used by permission]

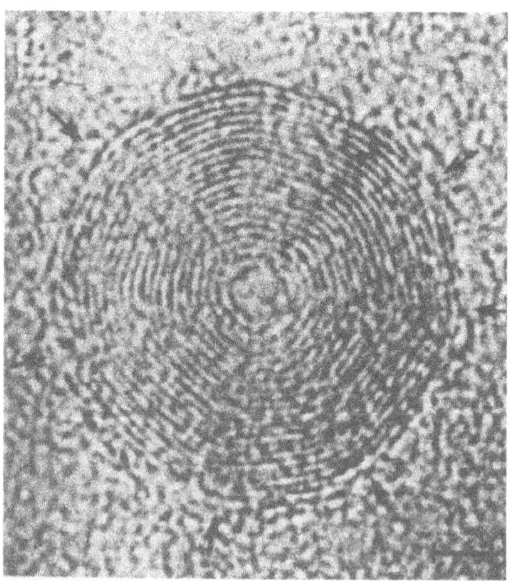

polyhedral particles, the pentagons in the nested shells are all aligned, making it possible for the graphene sheets to facet, while on the other hand, in the spherical onions, the pentagons are randomly oriented in the nested shells.[28] Interestingly, Ugarte pointed to the possibility of forming macroscopic graphite spheres by adequate annealing of the carbon. Indeed, macroscopic spheres of graphite already exist. The best-known examples of macroscopic graphite spheres are those in certain cast irons. Less well known are those that occur naturally in sizes ranging from a few microns in diameter to 2 cm in diameter in igneous and metamorphic rocks.

It has long been known that spheres of graphite, typically on the order of 0.1 mm in diameter, occur in certain cast irons that have superior mechanical properties compared to other cast irons.[43] Since the spheroidal graphite is important in controlling the properties of cast iron, the nucleation, growth, and microstructure of such graphite have been studied intensely and reported in the metallurgical literature.[44] The production of cast iron containing spherical graphite, as opposed to flake graphite, usually requires the presence of "nodularizers" such as Ce or Ce and Mg, which seem to promote the nucleation and growth of nodular (spherical) graphite. However, Sadocha and Gruzleski have suggested that in Fe-C-Si alloys the "natural" morphology for graphite is spheroidal instead of tabular, and the nodularizers merely act as scavengers to remove deleterious impurities.[45]

Similar to the carbon onions (Figure 6-5) observed by Ugarte[27] and Iijima[13,42] (compare also heat-treated CCVD fibers, as in Figure 6-3b), the spherical graphite in cast iron also has its graphene sheets aligned circumferentially, despite a radial-appearing texture.[46-48] In fact, Miao et al. supposed that the nucleus of spherical graphite in cast iron might be a C_{60} polyhedron.[48] Double and Hellawell noted that very thin graphite sheets should be able to roll up easily into loose spheres while floating freely in solution due to their extreme flexibility. Further, they proposed that the radial texture and the circumferential alignment of the graphene sheets result from a cone-helix growth mechanism from a common center (Figure 6-4).[14-16]

Because of its relation to the metallurgical formation of spherical graphite, it is interesting to note here the occurrence of a type of spherulitic graphite, known as *cliftonite,* which occurs with diameters ranging from 10 μm to 1 mm in some iron meteorites.[49,50] Cliftonite is a polycrystalline aggregate of graphite that most commonly shows a cubic or cubo-octahedral morphology. The morphology can also be dodecahedral, cubododecahedral, or spherical. In cube or cubo-octahedral grains, the graphite crystallites radiate out from the centers of the grains to the morphological cube or octahedron faces, respectively, with the graphene planes of the graphite aligned approximately parallel to these respective faces. In light of the work of Sadocha and Gruzleski,[45] it is interesting to note that cliftonite is composed of carbon of exceptionally high purity. It is believed that cliftonite grows by the decomposition of cohenite [naturally occurring $(Fe,Ni,Co)_3C$] in kamacite [naturally occurring (Fe,Ni)], with subsequent diffusion of the carbon to nucleation sites, where spherulitic growth takes place but where the external morphology is controlled by the surrounding metal.[50]

Another interesting occurrence of spherical graphite is in the quenched regions of diamond and graphite melted in a diamond anvil with a pulsed YAG laser.[51] In these quenched melts, spherical particles were observed with a range of diameters. In spheres up to 0.2 μm in diameter, the particles were graphitic, and the graphene sheets were found to be concentric. On the other hand, spheres of diameter greater than 0.2 μm were shown to be polycrystalline diamond with either a radial or a granular texture. We note also that a rare form of diamond, known as *bort* or *ballas*, occurs naturally, with a spherical morphology and a radial internal texture.[52] The origin and microstructure of these spherical diamonds do not seem to have been systematically investigated, though they could possibly have important commercial applications. Both diamond and graphite have been found in graphite/nickel composite powders subjected to high-pressure shock-wave compression.[53] Here, carbon spheres up to 7 μm in diameter occur as solid graphite or as thin graphite rings that sometimes surround diamond. Evidence from these samples supports a growth mechanism[54] whereby the graphite crystallizes on cooling on the periphery of gas bubbles in the metallic melt, and crystallization proceeds inward in a winding fashion, with the graphene sheets oriented normal to the spheres' radii. A graphitic carbon sphere 65 nm in diameter has been produced by shock compression of carbon black,[55] and shows an onion-like structure in TEM similar to those made by Ugarte.

Regarding *naturally* occurring spherical carbons, the fullerenes C_{60} and C_{70} have been discovered as in the carbonaceous rock shungite from Shunga, Karelia, Russia[56] and in a fulgerite (a glassy, tubular structure resulting from lightning striking the ground) from Sheep Mountain, Colorado.[57] On a larger length scale, it has recently been pointed out[58] that graphitic particles, which may be related to carbon onions, had been observed occurring up to 500 Å in diameter in the Allende meteorite (a type of carbon-rich meteorite known as a *carbonaceous chondrite,* which fell in Chihuahua, Mexico, in 1969).[59] Spherical graphite particles, which are described as being of interstellar origin based on their anomalous

^{12}C:^{13}C ratios, occurring up to several microns across, have been isolated from the Murchison meteorite, a carbonaceous chondrite that fell in Victoria, Australia, in 1969.[60-62] Graphites of other morphologies also occur in this meteorite, but only the spherical ones show the anomalous carbon isotope ratios.[63] The interstellar spherical graphite particles occur as two types. The first shows smooth surfaces, sometimes with shell-like platy structures; the second consists of a dense aggregate of small scales. TEM has revealed that both types have concentric-layered internal structures; the first type has layers of well-crystallized graphite, and the second type has smaller scales with a lower degree of crystallinity.[64] The second type also shows a hint of a radial texture in the more disordered regions (Figure 6-6). To date, however, simple fullerenes have not been found in carbonaceous chondrites.[65]

Perhaps as a footnote to the question of "The ultimate fullerene?", it is of interest to note that spheres of graphite occur naturally in both metamorphic and igneous rocks, in sizes ranging from submillimeter to over 2 cm in diameter. In contrast to those occurring in cast iron, naturally occurring spherical graphites have received relatively little attention.

The surface[44-47] and internal[15,46,47] textures of the natural graphite spheres are remarkably similar to those occurring in cast iron, despite their obviously very different origins. For example, the smooth surfaces on the 9-mm graphite spheres from Brazil shown in Figure 6-7 are very much like those illustrated by Lux in Ref. 44. The internal texture of graphite spheres in cast iron is typically radial, as shown in Figure 3 of Ref. 15, and is also radial in naturally occurring spheres, such as is shown in Figure 6-8 of a rather large sphere from Sussex County, New Jersey.

Kvasnitsa and Yatsenko[66] have performed the only studies to date focusing in detail on the structure of natural spherical graphite, using samples from the Azov Sea area of the Ukrainian Shield. The spherical graphite from this area occurs up to 1.5 mm in diameter. These spheres often show a three-part structure consisting of a core of randomly oriented crystallites, a large middle section with a radial texture, and a thin surface shell of concentric texture (Figure 6-9). By means of X-ray diffraction they demonstrated that, similar to spherical graphite in cast iron, both the radial middle section and the concentric outer shell are composed of a concentric arrangement of graphite layers.

In the calcite marble at the quarry in Pargas, Finland, a large variety of graphite morphologies occur, including rosettes (Figure 6-10) and well-formed single crystals. In addition, tiny graphite spheres (Figure 6-11) occur up to 1 mm that are strikingly similar to those illustrated by Hunter and Chadwick.[47] The surfaces of some of these spheres show

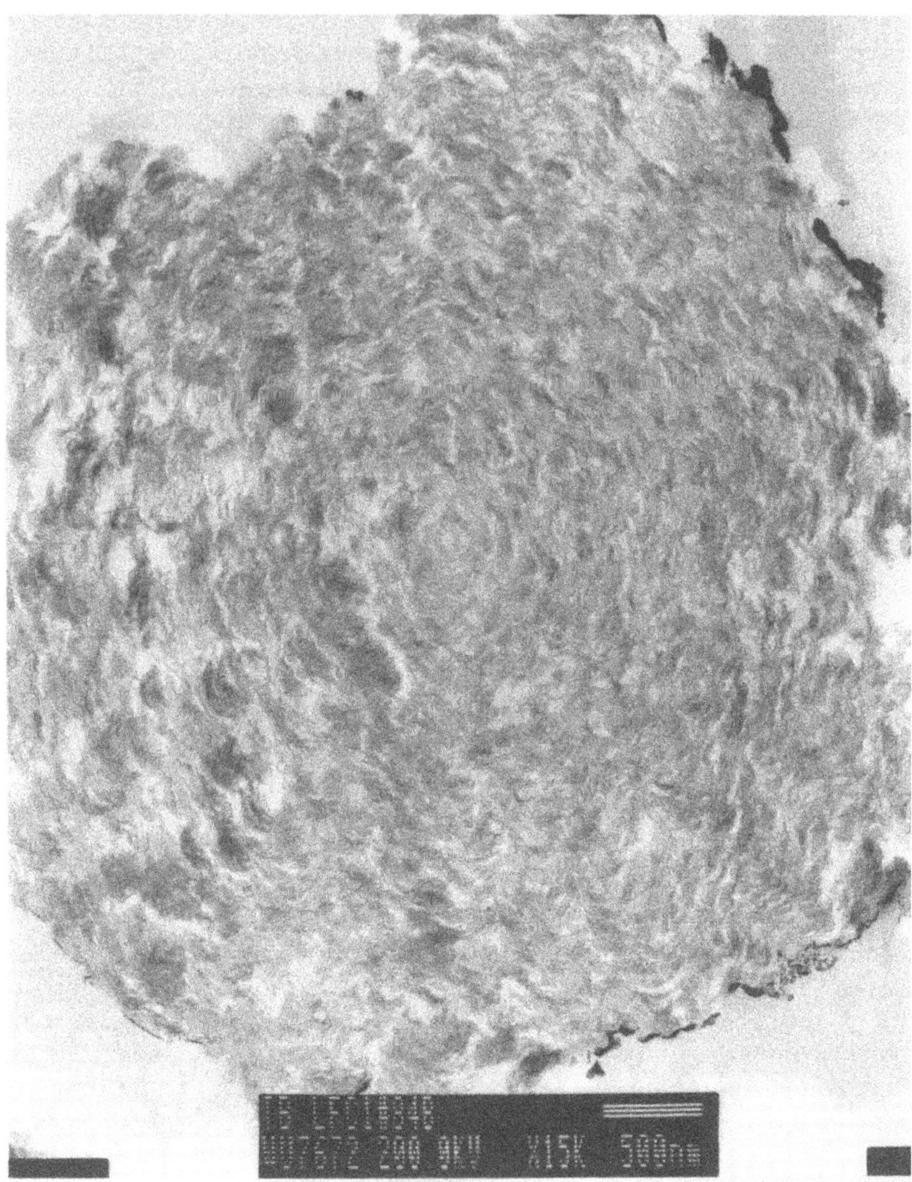

FIGURE 6-6 TEM bright-field micrograph of a 70-nm, nearly central, thin section of a scaly, interstellar graphite sphere from the Murchison meteorite. (Photo courtesy of T.J. Bernatowicz[64])

FIGURE 6-7 Group of three graphite spheres (9-mm diameter) from Arroyal de Bareiras, Brazil. Harvard Mineralogical Museum No. 105887.

hexagonal outlines of individual graphite crystals, with their basal planes lying parallel to the surface, which suggests that the internal structure of the spheres is a circumferential arrangement of the graphene layers. Again, broken spheres from Pargas show a radial interior texture.

A continuing examination of private and public mineral collections, and a thorough search of the literature, has revealed that spherical graphite is found in a relatively large number of natural occurrences, currently over 20. Numerous localities for graphite spheres occur in the marbles in several formations in the large region extending from New Jersey through New York and on into Ontario, Canada. Some of the largest graphite spheres known seem to occur in this region. The Franklin and Sterling Hill mines in Sussex County, New Jersey, are particularly noteworthy,[67] as spheres of graphite are known to occur there (Figure 6-8) up to 2 cm in diameter, although 1 cm or less is more typical. Gooderham and Minden, Ontario, also have graphite spheres up to 5 mm in diameter in calcite. Some of the igneous occurrences include the Gillette Quarry, Haddam Neck, Connecticut; the Henshaw quarry, Riverside County, California; and a pegmatite outcrop north of Custer, South Dakota. These localities produce graphite spheres up to 1 mm across. Clearly, it will be

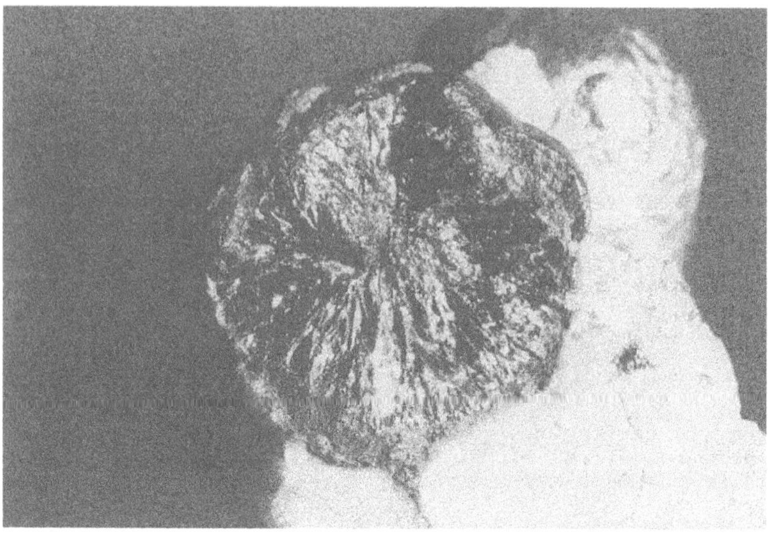

FIGURE 6-8 Cross section of a 1.5-cm broken graphite sphere in calcite from Sussex County, New Jersey, illustrating the radial interior texture reminiscent of spheroidal graphite in cast iron. Compare with Figure 3 in Ref. 15. Seaman Mineral Museum specimen No. DM1504.

FIGURE 6-9 Sketch of the typical interior texture of graphite spheres from the Azov Sea area of the Ukrainian Shield. Region I is an aggregate of randomly oriented crystallites. The graphite layers are circumferential in both regions II and III. (After Kvasnitsa and Yatsenko.[66])

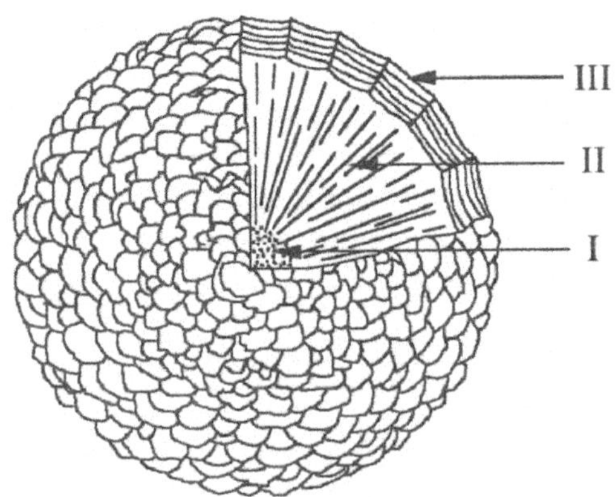

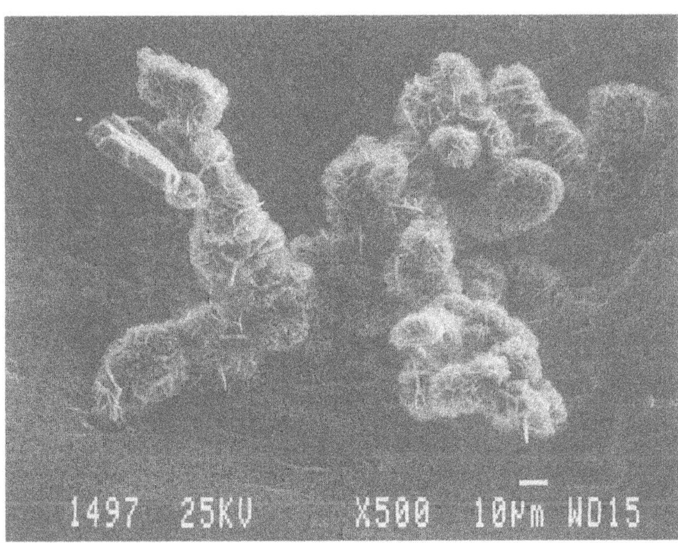

FIGURE 6-10 Aggregate of graphite rosettes in HCl-etched calcite from Pargas, Finland. Such rosettes resemble flake graphite in cast iron and Ni-C alloys. Compare with Figure 4 in Ref. 47, which shows a 0.1-mm graphite rosette from an etched, furnace-cooled Ni-C alloy. JAJ specimen No. 1394.

of interest to study the structures and geological origins of these natural graphite spheres in greater detail.

Micron-sized spheres of graphite occur in a natural coke at St. Ingbert mine, Saar basin, Germany.[68] Here the coke was formed by an igneous intrusion, of approximately 800 °C, into a coal deposit. The spherical graphites are supposed to have formed in the porous voids in the coke from the vapor phase. One may wonder if a modern examination of this material might yield fullerenes as well.

Finally, it is interesting to note that nested, fullerene-like structures similar to carbon onions have recently been reported[69] in deposited MoS_2 (a layered hexagonal material with many physical and structural similarities to graphite). It appears that curved growth can take place around a molybdenum vacancy where three or four sulfur hexagons join to form a triangle or rhombus. Remarkably, although molybdenite (natural MoS_2) usually forms tabular hexagonal crystals, both spherical and scroll-like whisker morphologies (Figure 6-12), all very reminiscent of graphite morphologies, have been observed up to 1 mm in diameter from the quarries at Mont Saint-Hilaire, Quebec, Canada.[70] The origin and the detailed structures of both cylindrical and spherical natural molybdenite remain to be investigated.

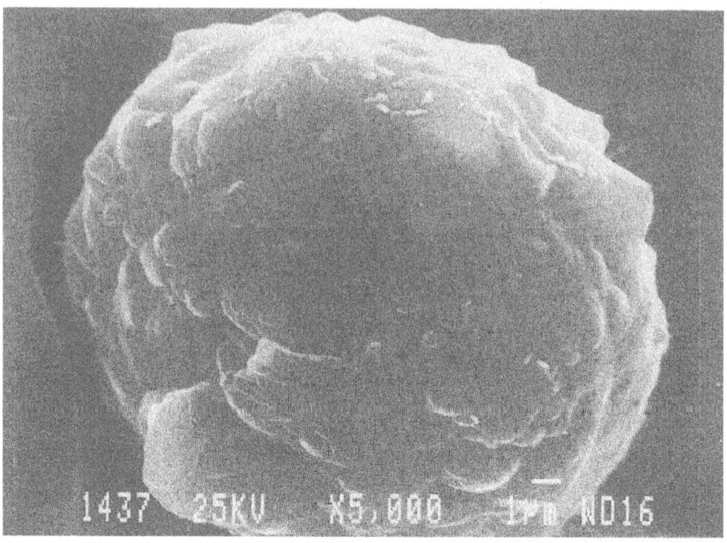

FIGURE 6-11 Graphite sphere (18 μm) in HCl-etched calcite from Pargas, Finland, showing hexagonal outlines of individual graphite crystals, with their basal planes lying parallel to the surface of the sphere. Compare with Figure 8 in Ref. 47, which shows a 0.06-mm graphite sphere of similar morphology from an etched, furnace-cooled Fe-C-Si alloy. JAJ specimen No. 1394.

The large number of natural occurrences, both igneous and metamorphic, of spherical graphite, as well as the wide variety of artificial conditions in which they can be produced, give additional strength to the notion that planar graphite may not be the most stable allotrope of carbon in systems of limited size.

Miao et al.[71] have recently reported that graphite platelets in spherical graphite from cast iron have a rhombohedral structure,[72] while interplatelet areas are mostly hexagonal but randomly oriented graphite. They have proposed a modified cone-helix growth mechanism[14–16] with $\alpha = 180°$.

Two new geologic sources of the fullerenes C_{60} and C_{70} have been reported: the Onaping Formation of the Sudbury impact structure, Ontario, Canada,[73] and two Cretaceous-Tertiary boundary sites in New Zealand.[74]

Acknowledgments

I am grateful to William Kelly and Michael Hansen of the New York State Museum; George Robinson and Louis Moyd of the Canadian Museum of Nature; Jamshid Hassanzadeh of the Department of Earth and Space Sciences at the University of California, Los Angeles; and Carl Francis of the Harvard Mineralogical Museum for making samples available for study and photography. I am also grateful to Kimmo Pietikäinen, Edwin L. Clop-

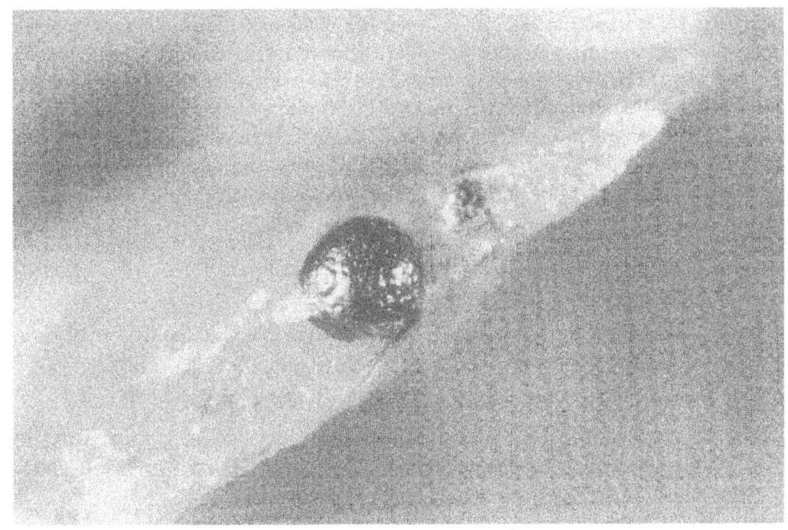

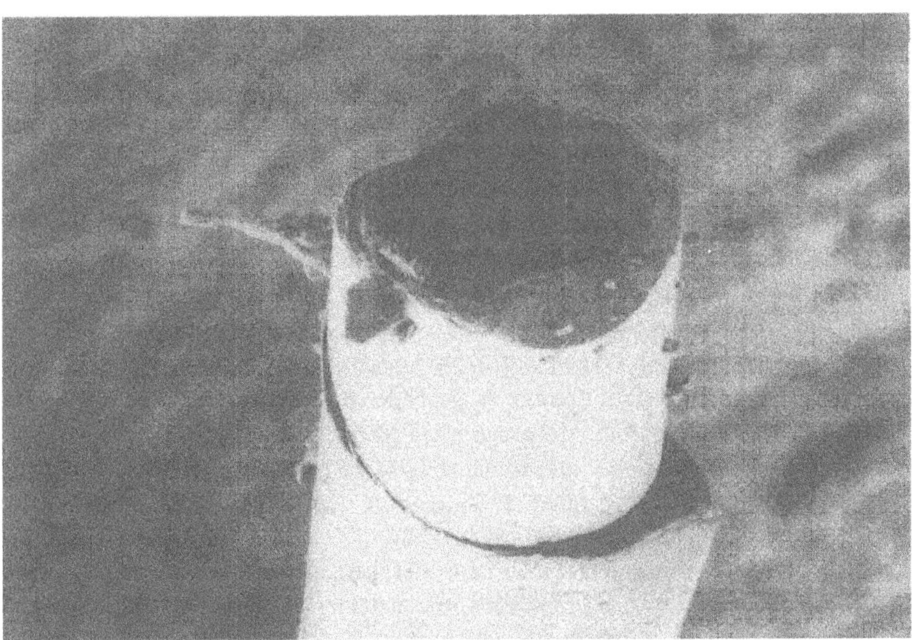

FIGURE 6-12 Molybdenite from Mont Saint-Hilaire, Quebec, Canada, showing rare curved morphologies. (a) A 0.5-mm sphere in sodalite. JAJ specimen No. 1358. (b) A 0.5-mm-diameter by 6-mm-long, scroll-type whisker. Canadian Museum of Nature specimen No. 48756. (SEM photo courtesy of E. Vadas.)

ton, Vicky Bobofchak, and Marcelle Weber for supplying specimens. I thank Edward and Nina Nadgorny for translating Ref. 66. R. Peter Richards and Quintin Wight have provided helpful correspondence and samples of spherical molybdenite from Mont Saint-Hilaire. I have received several elusive references from Pete Dunn, Fred DeVito, Dana Morong, Victor Kvasnitsa, Richard Bostwick, Robert E. Jenkins II, Stephen Sanford, and Steve Misiur. I am indebted to Ruth Kramer and the Institute of Materials Processing at Michigan Technological University for supplying SEM photographs of graphite, and to László Horváth and Elizabeth Vadas for the SEM photograph of tubular molybdenite.

REFERENCES

1. M.S. Dresselhaus, G. Dresselhaus, K. Surihara, I.L. Spain, and H.A. Goldberg, *Graphite Fibers and Filaments* (Springer, Berlin, 1988).
2. E.J. Freise, *Nature* (London), **193,** 671 (1962).
3. For a review of the relations between crystal structure and external shapes see, for example, *Crystal Form and Structure,* C.J. Schneer, Ed. (Dowden, Hutchison & Ross, Stroudsburg, PA, 1977).
4. J.A. Jaszczak, *Mineralogical Record,* **21,** 427 (1991).
5. A.R. Ubbelhode, *Endevour,* **24**(92), 63 (1965).
6. C. Palache, *Am. Mineral,* **26,** 709 (1941).
7. S.B. Austerman, in *Chemistry and Physics of Carbon,* Vol. **4,** P.L. Walker, Jr., Ed. (Marcel Dekker, New York, 1968), p. 137.
8. For a review of less ordered spherical carbons in carbon black, see A. Oberlin, in *Chemistry and Physics of Carbon,* Vol. 22, P.A. Thrower, Ed. (Marcel Dekker, New York, 1989), p. 1.
9. H.W. Kroto, *Nature* (London), **329,** 529 (1987).
10. S.E. Stein and R. Leonard, *Mol. Struct. Energ,* **2,** 37 (1987).
11. D. Ugarte, *Chem. Phys. Lett.,* **198,** 596 (1992).
12. H.W. Kroto, *J. Chem. Soc. Dalton Trans.,* **10,** 2141 (1992).
13. H.W. Kroto and K. McKay, *Nature* (London), **331,** 328 (1988).
14. D.D. Double and A. Hellawell, in *The Metallurgy of Cast Iron,* B. Lux, I. Minkoff, and F. Mollard, Eds. (Georgi, St. Saphorin, Switzerland, 1975), p. 509.
15. D.D. Double and A. Hellawell, *Acta Metall.,* **22,** 481 (1974).
16. D.D. Double and A. Hellawell, *Acta Metall.,* **17,** 1071 (1969).
17. S. Amelinckx, W. Luyten, T. Krekels, G. Van Tendeloo, and J. Van Landuyt, *J. Cryst. Growth,* **121,** 543 (1992).
18. There are numerous reviews on both the history and science of fullerenes. See, for example, R.E. Smalley, *The Sciences,* **31,** 22 (1991); R.F. Curl and R.E. Smalley, *Sci. Am.,* **265,** 54 (1991); *Phil. Trans.: Phys. Sci. Eng.,* **343,** 1 (1993) (*A postbuckminsterfullerene view of the chemistry, physics and astrophysics of carbon.* A discussion organized and edited by H.W. Kroto, A.L. MacKay, G. Turner, and D.R.M. Walton); R.C. Haddon, *Science,* **261,** 1545 (1993); M.S. Dresselhaus, G. Dresselhaus, and P.C. Eklund, *J. Mater. Res.,*

8, 2054 (1993); *The Fullerenes,* H.W. Kroto, J.E. Fischer, and D.E. Cox, Eds. (Pergamon, Oxford, 1993); *Buckminsterfullerenes,* W.E. Billups and M.A. Ciufolini, Eds. (VCH, New York, 1993).

19. M. Hillert and N. Lange, *Z. Krist.,* **111,** 24 (1958).
20. R. Bacon, *J. Appl. Phys.,* **31,** 283 (1960).
21. R.T.K. Baker and P.S. Harris, in *Chemistry and Physics of Carbon,* Vol. 14., P.L. Walker, Jr., and P.A. Thrower, Eds. (Marcel Dekker, New York, 1978), p. 83.
22. G.G. Tibbetts, in *Carbon Fibers, Filaments, and Composites,* J.L. Figueiredo et al., Eds. (Kluwer, Dordrecht, 1990), p. 73.
23. N.M. Rodriguez, *J. Mater. Res.,* **8,** 3233 (1993).
24. A. Yoshida, Y. Hishiyama, and M. Endo, in *Extended Abstracts of the 17th Biennial Conference on Carbon, Lexington, KY* (American Carbon Society, University Park, PA, 1985), p. 297.
25. S. Iijima, *Nature* (London), **354,** 56 (1991).
26. S. Iijima, P.M. Ajayan, and T. Ichihashi, *Phys. Rev. Lett.,* **69,** 3100 (1992).
27. D. Ugarte, *Nature* (London), **359,** 707 (1992).
28. D. Ugarte, *Europhys. Lett.,* **22,** 45 (1993).
29. D. Ugarte, *Chem. Phys. Lett.,* **207,** 473 (1993).
30. P.M. Ajayan and S. Iijima, *Nature* (London), **361,** 333 (1993).
31. S. Iijima and T. Ichihashi, *Nature* (London), **363,** 603 (1993).
32. D.S. Bethune, C.H. Klang, M.S. de Vries, G. Gorman, R. Savoy, J. Vazquez, and R. Beyers, *Nature* (London), **363,** 605 (1993).
33. A.R. Patel and S.V. Deshapande, *Carbon,* **8,** 242 (1970).
34. H. Murayama and T. Maeda, *Nature* (London), **345,** 791 (1990).
35. W. Krätschmer, L.D. Lamb, K. Fostiropoulos, and D.R. Huffman, *Nature* (London), **347,** 354 (1990).
36. R.C. Haddon, A.F. Hebard, M.J. Rosseinsky, D.W. Murphy, S.J. Duclos, K.B. Lyons, B. Miller, J.M. Rosamilia, R.M. Fleming, A.R. Kortan, S.H. Glarum, A.V. Makhija, A.J. Muller, R.H. Eick, S.M. Zahurak, R. Tycko, G. Dabbogh, and F.A. Thiel, *Nature* (London), **350,** 320 (1991).
37. R.M. Fleming, T. Siegrist, P.M. March, B. Hessen, A.R. Kortan, D.W. Murphy, R.C. Haddon, R. Tycko, G. Dabbogh, A.M. Mujsce, M.L. Kaplan, and S.J. Zahurak, *Mat. Res. Soc. Symp. Proc.,* **206,** 691 (1991).
38. P.A. Heiney, J.E. Fischer, A.R. McGhie, W.J. Romanow, A.M. Denestein, J.P. McCauley, Jr., A.B. Smith III, and D.E. Cox, *Phys. Rev. Lett.,* **66,** 2911 (1991).
39. R.L. Meng, D. Ramirez, X. Jiang, P.C. Chow, C. Diaz, K. Matsuishi, S.C. Moss, P.H. Hor, and C.W. Chu, *Appl. Phys. Lett.,* **59,** 3402 (1992).
40. Y. Yosida, *Jpn. J. Appl. Phys.,* **31** Pt. 2, L505 (1992).
41. M. Kléman, in *Dislocations in Solids,* Vol. 5., F.R.N. Nabarro, Ed. (North-Holland, Amsterdam, 1980), p. 243.
42. S. Iijima, *J. Cryst. Growth,* **50,** 675 (1980).

43. H. Morrogh and W.J. Williams, *J. Iron Steel Inst.*, **155,** 321 (1947).
44. B. Lux, *AFS Cast Metals Res. J.*, **8,** 25 (1972); **8,** 49 (1972).
45. J.P. Sadocha and J.E. Gruzleski, in *The Metallurgy of Cast Iron*, B. Lux, I. Minkoff, and F. Mollard, Eds. (Georgi, St. Saphorin, Switzerland, 1975), p. 443.
46. M.J. Hunter and G.A. Chadwick, *J. Iron Steel Inst.*, **210,** 117 (1972).
47. M.J. Hunter and G.A. Chadwick, *J. Iron Steel Inst.*, **210,** 701 (1972).
48. B. Miao, K. Fang, W. Bian, and G. Liu, *Acta Metall. Mater.*, **38,** 2167 (1990).
49. R. Brett and G.T. Higgins, *Science*, **156,** 819 (1967).
50. R. Brett and G.T. Higgins, *Geochim. Cosmochim. Acta*, **33,** 1473 (1969).
51. M. Weathers and W.A. Bassett, *Phys. Chem. Minerals*, **15,** 105 (1987).
52. Y.L. Orlov, *The Mineralogy of the Diamond* (Wiley, New York, 1977), pp. 14-15.
53. I. Simonsen, S. Chevacharoenkul, Y. Horie, T. Akashi, and H. Sawaoka, *J. Mater. Sci.*, **24,** 1486 (1989).
54. H.H. Stadelmaier, *Z. Metallkunde*, **51,** 601 (1960).
55. K. Yamada, H. Kunishige, and A.B. Sawaoka, *Naturwissenschaften*, **78,** 450 (1991).
56. P.R. Buseck, S.J. Tsipursky, and R. Hettich, *Science*, **257,** 215 (1992).
57. T.K. Daly, P.R. Buseck, P. Williams, and C.F. Lewis, *Science*, **259,** 1599 (1993).
58. L. Becker, G.D. McDonald, and J.L. Bada, *Nature* (London), **361,** 595 (1993).
59. P.P.K. Smith and P. Buseck, *Science*, **212,** 322 (1981).
60. S. Amari, E. Anders, A. Virag, and E. Zinner, *Nature* (London), **345,** 238 (1990).
61. I. Wright, *Nature* (London), **365,** 786 (1993).
62. S. Amari, P. Hoppe, E. Zinner, and R.S. Lewis, *Nature* (London), **365,** 806 (1993).
61. E. Anders and E. Zinner, *Meteoritics*, **28,** 490 (1993).
64. T.J. Bernatowicz, S. Amari, E.K. Zinner, and R.S. Lewis, *Astrophys. J.*, **373,** L73 (1991).
65. M.S. de Vries, K. Reihs, H.R. Wendt, W.G. Golden, H.E. Hunziker, R. Flemming, E. Peterson, and S. Chang, *Geochim. Cosmochim. Acta*, **57,** 933 (1993).
66. V.N. Kvasnitsa and V.G. Yatsenko, *Mineralogicheskii Zhurnal*, **13**(1), 95 (1991).
67. C.S. Lemanski, Jr., *The Picking Table*, **32**(1), 11 (1991).
68. E. Stach, *Z. Deut. Geologischen Gesellschaft*, **103,** 233 (1951).
69. L. Margulis, G. Salitra, R. Tenne, and M. Talianker, *Nature* (London), **365,** 113 (1993).
70. L. Horváth and R.A. Gault, *Mineral. Record*, **21,** 284 (1990).
71. B. Miao, D.O. Northwood, W. Bian, K. Fang, and M.H. Fan, *J. Mater. Sci.*, **29,** 255 (1994).
72. F. Laves and Y. Baskin, Z. Krist., **107,** S.337–356 (1956).
73. L. Becker, J.L. Bada, R.E. Winans, J.E. Hunt, T.E. Bunch, and B.M. French, *Science* **265,** 642 (1994).
74. D. Heymann, L.P.F. Chibante, R.R. Brooks, W.S. Wolbach, and R.E. Smalley, *Science*, **265,** 645 (1994).

1

Fractal Index and Fractal Notation

G. David Mendenhall

FRACTAL INDEX

Large molecules of the arborol (see Chapter 2, this volume) or dendrimer type pose difficulties for both verbal and written communication. For instance, consider compound **5** (p. 29).[1]

Out of curiosity, we sent a letter with the structure to *Chemical Abstracts* in 1985 and requested an official name for it. The director of nomenclature replied that its preferred IUPAC name would be

> 3,3′-[[2-Pentyl-2-[[3,3,3-tris[[2-hydroxy-1,1-bis(hydroxymethyl)ethyl] carbamoyl]propoxy]methyl]trimethylene]-dioxy]bis[N,N′N″-tris [2-hydroxy-1,1-bis(hydroxymethyl)ethyl]-1,1,1-propanetricarboxamide]

This was also the *Chemical Abstracts* name for the 1967–71 index period. Since 1971 the name had been changed by *Chemical Abstracts* to

> 1,19-Dihydroxy-N,N′,N‴-tetrakis[2-hydroxy-1, 1-bis(hydroxymethyl)ethyl]-10-[[4-[[2-hydroxy-1, 1-bis(hydroxymethyl)ethyl]amino-3,3-bis[[[2-hydroxy-1,1-bis(hydroxymethyl)ethyl]amino]carbonyl]-4-oxobutoxy]methyl] -2,2,18,18-tetrakis(hydroxymethyl)-4-16-dioxo- 10-pentyl-8, 12-dioxa-3,17-diazanonadecane-5,5,15, 15-tetracarboxamide

The official names shown above are, except for specialists,[2] clearly useless for the purpose of constructing a structure from the name and require unreasonable effort to assign a name from the structure. Moreover, the time-dependence of the length of the official names indicates that the situation is not improving. In practice, articles in a journal, of course, are commonly written with structures drawn out and assigned simple numbers. This arrangement is satisfactory except for a written abstract or for verbal communication, in which one must continuously have the structures in view in order to relate the numbers to them. Moreover, some structures of the dendrimer type require nearly a page to display in readable form.

For these reasons, my graduate student Eric Chen and I began communicating in what we called *fractal notation* in connection with synthetic projects. The notation was described in a footnote[3,4] and is given here in greater detail. Multiply branched molecules with a high degree of symmetry could be described very concisely in the following general way:

(terminal group)$f_{(\text{outermost hub}\ldots\text{innermost hub})}$(n.n′.n″. . . n^z)core

The four parts of this notation are the terminal or peripheral group, the subscript, the connector groups, and the core.

The *terminal group* is left blank if it is H. For the [27]-arborol on page 29, which for the moment we will call 1,19-etc., the terminal group is HO. If a mixed functionality is present, such as a partially acetylated derivative, it may be designated $<X>_x<Y>_y$, where X and Y are the statistically average values of groups <X> and <Y>. The number of OH groups can be calculated with ease from the rest of the fractal notation, so it is redundant but still helpful to list the terminal group as $(HO)_{27}$. If the compound has a cyclic structure, the peripheral group is absent or incorporated into the core, and this is signified by a Ø in the terminal group position. (Confusion of Ø with the phenyl group may be avoided by abbreviating the latter as "Ph.")

The "f" stands for fractal and informs the reader that the notation under discussion is in effect, rather than some other encoding based on dates, personal names, or laboratory notebook page numbers.

The *subscript after the f* refers to the nature of the hubs (junctures) in the molecule. If the junctures are tetravalent C atoms, the subscript is omitted.

If the juncture is *always* the same atom (or molecular group), that atom or group is given only once as the subscript. If the junctures change as one traces the molecule from outward in, they are listed sequentially in the subscript in that order and separated by periods. For instance, the juncture portion of the molecule $[(CH_3)_3Si]_4C$ would be written "$f_{Si.C}$".

The *connector groups* in the highly branched molecule are listed from outside in, on the same line as the peripheral group and the f. They are enclosed in parentheses and separated by periods. For our compound 1,19-etc., the connector portion is thus (CH_2.NHCO.$CH_2CH_2OCH_2$). If one juncture in the molecule is bonded to the next juncture directly, the connector is written as a zero, Ø. For instance, the connector portion of the notation for $[(CH_3)_3Si]_4C$ would be written "(CH_2.Ø)".

The *core* is given without parentheses at the far right of the fractal notation. It is omitted if it is the same as the juncture atom or group and will become apparent from the subscript (or lack thereof) after the f. For 1,19-etc. the core is "C-C_5H_{11}-n," it being understood that three connector groups join the carbon to satisfy the normal valence state. If the core is listed only as zero, the bond that would be connected to it is connected to an identical subunit (i.e., the molecule is a dimer of the subunit).

The complete fractal notation for 1,19-etc. is then, according to our rules:

HOf(CH_2.NHCO.$CH_2CH_2OCH_2$)C-C_5H_{11}-n

We introduce one final simplification that is possible because the CH_2 group is the most common part group in such molecules. The *number* of consecutive methylene groups is substituted into the name, so that our compound becomes

HOf(1.NHCO.2O1)C5H

or, with redundant but useful detail

$(HO)_{27}$f(1.NHCO.2O1)C5H

The last two representations reveal a potential ambiguity between zero and an oxygen atom, which can be avoided by adding a slash to the former, as it is done with many computer printouts. Thus f_N(1O1) is $(CH_3OCH_2)_3N$, whereas f_N(1Ø1) is $(n\text{-}C_{101}H_{203})_3N$.

When writing out the chemical structure corresponding to a given fractal notation, it is helpful to note that for molecules with more than one "generation," the sequence of atoms and groups proceeds from peripheral group, to leftmost connector, to first juncture, second connector, second juncture, . . . to core. The order is shown with arrows in the following general formula:

$$Af_{M.N}(W.X.Y)Z \equiv [(\{[(A_rW)_sM]_tX\}_uN)_vY]_wZ$$

The subscripts r . . . w in the formula above are, of course, the number of connectors joined to each respective hub atom or group. They are omitted in fractal notation but can be inferred from the normal valence (strictly, the coordination number) of the atom or group. In those cases where it would be ambiguous because of multiple valency, it may be specified parenthetically with Arabic numerals (Roman numerals might result in confusion of "I" with iodine, and numbers without parentheses could be interpreted as CH_2 units). For instance, Tri-*tert*-butylphosphine and pentaethoxyphosphorane may be written $f_{C.P(3)}(1.\varnothing)$ and $f_{P(5)}(2O)$, respectively. (These two examples are given for illustration purposes; the usual formulas for them, of course, are simple enough to be preferred.)

With aromatic rings as connectors, it is preferable to use hyphens and o-, m-, and p- designators to avoid confusing numerical position indicators with numbers of methylene units. However, when the aromatic ring is not symmetrical, numbers become necessary. The numerical positions of connection are given with hyphens before and after the connector in the logical way. Thus the dendrimer **34** shown on p. 127 may be described as $[6\text{-}(2\text{-pyridone})]_4f(C{\equiv}C\text{-}3\text{-}(2\text{-pyridone})\text{-}6\text{-}C{\equiv}C\text{-p-}C_6H_4)$.

Fractal notation is very convenient for large molecules with a high degree of symmetry, which themselves are convenient to synthesize by iterated divergent or convergent sequences. When the molecule becomes highly irregular, the notation can become very complicated and offers little advantage over a conventionally written structure. However, since there are molecules of intermediate complexity, we illustrate fractal notation for different groups attached to a highlighted methine core with the following triglyceride analogue[5]:

$CH_2OCOCH_2OCH_2(CH_2OCH_2)_3CH_2OCH_3$
|
$\mathbf{CH}OCOCH_2OCH_2(CH_2OCH_2)_3CH_2OCH_3$
|
$CH_2OCO(CH_2)_{15}CH_3$

$f((1O1)_5CO_21)((1O1)_5CO_2)(16CO_21)\mathbf{CH}$

Each connector branch is enclosed in a separate set of parentheses. Similarly, a dendrimer (Figure 1F) that has unique connectors at one branch point, may be described in a way that describes the outer part with fractal notation as the peripheral substituent of the inner part:

$HOf_N(2NHCO2.6NHCO2)\ (2NHCO2.)_2f_N(2NHCO1)$

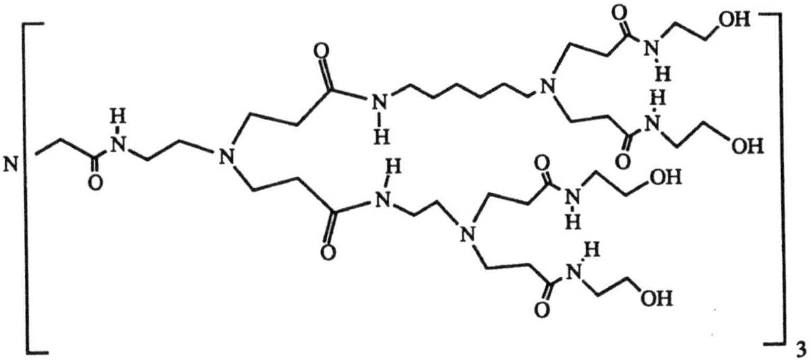

FIGURE 1F A hypothetical, asymmetric dendrimer.

Fractal notation is intended to be flexible. The point at which the different connector has been inserted can be made a little more obvious in this example by rewriting the outermost connector, which is the same for all branches of the molecule, as part of the peripheral group:

$HO2NHCO2f_N(Ø.6NHCO2)(Ø.2NHCO2)f_N(2NHCO1)$

Another simplification illustrated in several of the notations above is the use of parentheses and a subscript, instead of periods, to describe repeating groups. This approach vastly simplifies the notation for dendritic amines synthesized by Tomalia and coworkers.[6] For instance, the large dendrimer

$N[CH_2CH_2CONHCH_2CH_2N(CH_2CH_2CONHCH_2CH_2N$
$[CH_2CH_2CONHCH_2CH_2N(CH_2CH_2CONHCH_2CH_2NH_2)_2]_2)_2]_3$

can be described simply and unambiguously as $H_2Nf_N(2NHCO2.)_4$. The period within the parentheses signifies that four generations of connectors are involved, rather than one sequence repeated four times within a single layer of connectors.

An even larger molecule in this series, $H_2Nf_N(2NHCO2.)_52$, on inspection of its fractal notation, can immediately be seen to have a core derived from ethylenediamine and completely uniform, "fifth-generation" branching. The conventionally written formula requires careful scrutiny to arrive at these conclusions.

The fractal notation is simple enough to be useful for describing reaction of dendrimers, for example, for the last step in the synthesis of the largest known hydrocarbon, reported recently by Moore[7]:

$$(3,5\text{-di-t-butyl-}C_6H_3)_{16}f_{sym\text{-}C6H3}(C{\equiv}C.)_4C{\equiv}CH \xrightarrow[\text{sym-}C_6H_3I_3]{\text{Pd(II)/CuI} \atop Ph_3P/Et_3N}$$

$$(3,5\text{-di-t-butyl-}C_6H_3)_{48}f_{sym\text{-}C6H3}(C{\equiv}C.)_5$$

A stick-drawn representation of the product (Figure 2F) is admittedly more pleasing aesthetically.

Structures with multiply bonded junctures sometimes may be described with simple modifications of fractal notation. For instance, Shahlai's supertriptycene,[8] shown in Figure 3F, contains symmetrical methine pairs as junctures and benzene rings as the connectors. If we consider the outer benzene rings as hexasubstituted C_6 connectors each with four peripheral H atoms, a rather simple fractal notation can be written:

$f_{CH/CH}(1,2{:}3,4{:}5,6\text{-}C_6\text{-c.})_2$

In this example, the C_6-c defines a ring of six carbon atoms.

Fractal notation is rather easily assigned to simple monocyclic compounds. With the peripheral Ø defined earlier, for instance, 1-methylcyclooctane becomes Ø$f_{CHMe}(7)$.

Although fractal notation is usually straightforward, it is not intended to be a rigid exercise and one should always look for simplifications. For instance, the cascadol **10** (p. 35) may be written according to our rules as

$AcOf(2O2O1.(1O1)_5)$

On inspection, it can be seen that if one methylene is added to the peripheral group, we can condense the notation further:

$AcOCH_2f(1O1)_2.(1O1)_5$

or

$AcO1f(1O1)_{2.5}$

In the first of the two examples immediately above, to avoid some clutter, we dispensed with one normally proscribed set of parentheses around the connecting groups. We would not do this if it had been nec-

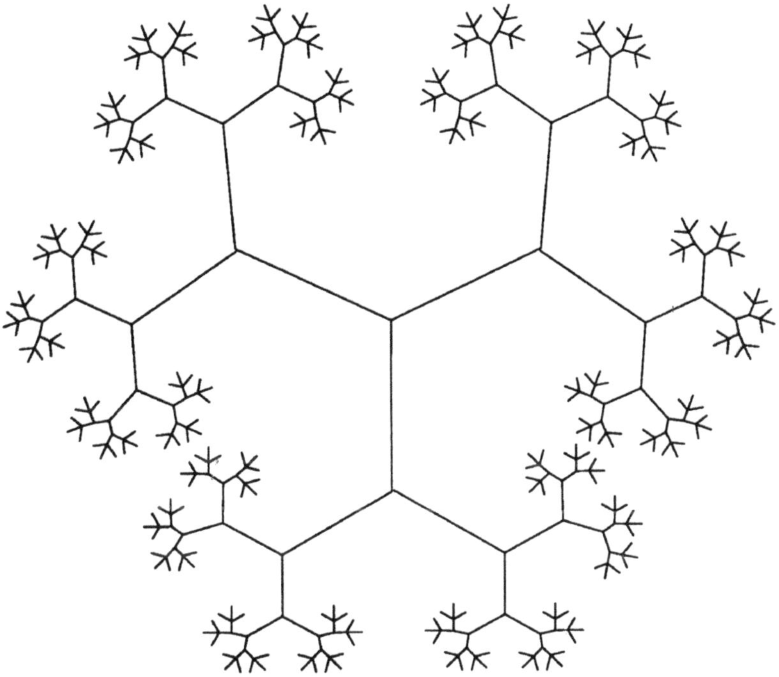

FIGURE 2F One of Moore's dendrimers. [From Z. Xu and J.S. Moore, *Ang. Chem. Int. Ed.*, **32,** 246 (1993)]

FIGURE 3F Supertriptycene. [From K. Shahlai and H. Hart, *J. Am. Chem. Soc.*, **112,** 3687 (1990)]

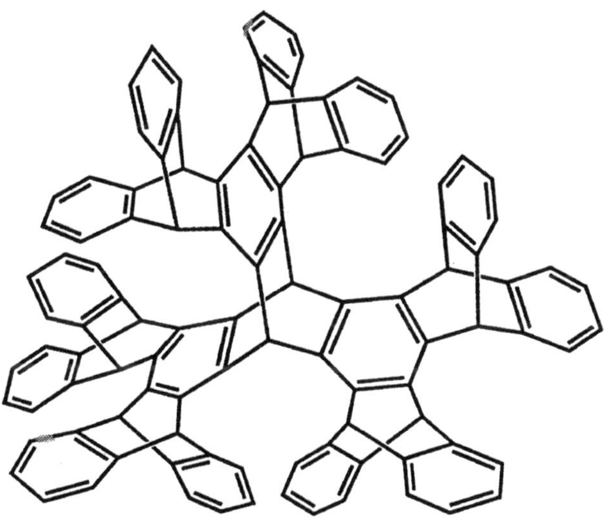

essary to write out the core explicitly, which could render the simplification ambiguous.

When constructing the fractal index that appears at the end of this book, it was apparent that many of the molecules given in the chapters were too intricate to be reduced to a simple line of code. New rules would have to be invented, and in the end the people at *Chemical Abstracts* might have a simpler name after all. Accordingly, only the fractal notations that were easily written with the rules given above were included in the index. In a few cases we compromised, because the dendrimer shape of the molecule was obvious in spite of the complexity. Thus the organometallic complex **10A** prepared by Balzani and coworkers (p. 89) appears in the index as the incompletely defined notation $(bpy)_{12}f_{Ru(3)}$(Ø.2,3-dpp.2,3-dpp).

Recently, a novel hydrocarbon material was synthesized from Na/K alloy and benzotrichloride[9]:

$$PhCCl_3 + 3Na/K \rightarrow (Ph\text{-}C)_n + 3Na/KCl$$

The structure of the resulting solid was inferred to be diamond-like, that is, one on which the benzylic carbons were joined by single covalent bonds. If one imagines successive Ph-C units branching off from a Ph-C core, one will be *tempted* to write the fractal notation as $f_{PhC}(Ø.)_n$, while diamond itself *might* be $f(Ø.)_n$. The fractal notation in this case gives more information than the alternative formula $(PhC)_n$, which could also designate linear poly(diphenylacetylene).

However, these representations are inadequate because they do not reveal the network connectivity between generations of C atoms surrounding the arbitrarily selected core atom. We cannot simply add a symbol representing a network without additional information to distinguish between possible geometrical forms of the lattices. The crystallographic unit cell designations are not suitable for this purpose because the cell boundaries often do not correspond to our "elementary" notion of the basic structural unit.

In some cases it may be possible to specify the elementary form of the network parenthetically as a prefix, e.g. <adamantane>$f(Ø.)_n$ for diamond. However, this practice may obscure unknown, real structural features of the material as in the case of the product from benzotrichloride.

On the other hand, the extremely large, polycyclic hydrocarbons of Moore[10,11] may be described very easily by incorporating the conventional nomenclature for small multicyclic rings into the fractal notation. The geometrical form is described in brackets in the prefix, with the numbers indicating—instead of numbers of carbons—how many repeating connector units link the junctures. The juncture group(s), and the structure of one (unit) connector are given in the usual way as a subscript after the f, and on the same line as the f, respectively.

Examples of this notation applied to a simple bicyclic hydrocarbon, and to a larger unit with the same shape are given in Figure 4F. A conventional name for the aliphatic hydrocarbons, and an alternate notation for each based on the approach proposed for supertriptycene is also included. For these two examples the latter is in fact the simplest, although this may not always be the case.

Fractal notation should lend itself easily to computer-assisted drawing of structures from the names.

In summary, fractal notation should assist chemists in a number of ways to communicate in terms of complicated structures, especially of the dendrimer and arboreal type.[5] When one of my students announced that he had succeeded in making "f one two C OH," I knew at once without resort to chalk and blackboard that he had prepared (t-Bu-$CH_2CH_2)_3COH$.

SUMMARY OF FRACTAL NOTATION

Molecules in this book that are easily named according to the rules of fractal notation are listed here in approximate order of increasing complexity. For some of the more complicated examples, the names do not describe the structure unambiguously without additional tedious detail. In those cases, it was felt that the names still convey sufficient useful information to warrant their inclusion in the table.

Summary of fractal notation:

$$\text{(peripheral groups)}_{\text{multiplicities}}\text{f}_{\text{junctures (coordination number)}}\text{(connectors)core}$$

1. The structure is described from the outside toward the core.
2. Consecutive *N* methylene units are abbreviated as *N*.
3. The peripheral group is omitted if it is H. (Otherwise "Hf" = hafnium!)
4. The juncture is omitted if all are tetravalent C in a fully symmetrical structure. An unusual isotopic designation is given in parentheses at the left of the juncture; the valence, if needed, is given in parentheses at the right.
5. The core is omitted if it is the same as or conveniently represented in the juncture. Tihexadecylcarbinyl chloride would be designated with an explicit core as f(16)CCl, or without one as $f_{CCl}(16)$ or $f(16)_3(Cl)$. The core may be univalent as for example n-hexadecyl chloride = f(16)Cl.

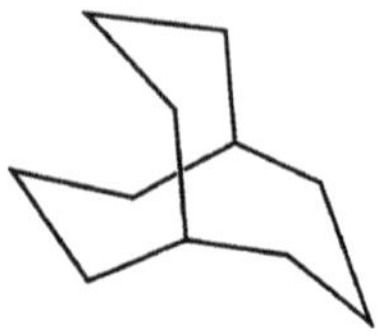

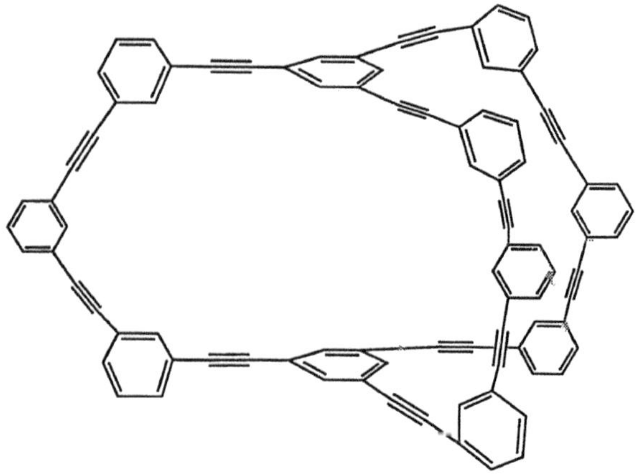

bicyclo[3.3.3]undecane

<bicyclo[3.3.3]>$f_{CH}(1)$ <bicyclo[3.3.3]>$f_{sym\text{-}C6H3C3}(\equiv C\text{-}m\text{-}C_6H_4\text{-}C\equiv)$

$f_{CH/CH}(3)$ $f_{sym\text{-}C6H3/sym\text{-}C6H3}C_2(\text{-}m\text{-}C_6H_4\text{-}C_2)_3$

FIGURE 4F Conventional and fractal nomenclature for multicyclic hydrocarbons.

6. Separation of different generations is designated by periods.
7. The use of Ø in connector indicates a direct bond from a peripheral group to a juncture or from juncture to juncture; Ø in the core indicates a dimer; Ø in a peripheral position indicates a cyclic or spiro molecule, as for example spiropentane, Øf(2)(2); "Ø" as subscript after the f indicates a cyclic structure defined in the connector only, so that cyclododecane is $f_{Ø}(12)$.
8. The formula for a ring as a core or hub has "-c" except for the conventionally described aromatic derivatives, such as *o*-, *m*-, and *p*-C_6H_4, sym-C_6H_3, and sym-C_6H_2. Asymmetric aromatic connectors, unless relatively simple (e.g., 2-pyridyl), may be described as "-position-(name of aromatic)-position-," where the positions are conventional numbers defining the points of attachments within the aromatic system.
9. Asymmetric structures connected to a hub are described as "f(branch1) (branch2) . . ." or as "f(branch1) (branch2) . . . f(x.y . . .)." If successive generations of connector groups are comprised of different numbers (x and y) of an otherwise identical repeating sequence, the connectors are abbreviated as "$(\text{sequence})_{x.y}$".
10. A lattice network with an identifiable, repeating elementary geometrical figure may be described in simple cases as "<geometrical form>f_{hub}(connector)," where the geometrical form may be "<cube>,"

"<adamantane>," and so on and the core designation, normally at the far right, is omitted.

11. Bicyclic compounds can be described as $f_{A/B}$(connector1) (connector2) . . . , where the connectors are described in the usual way from the bridgehead juncture group A to bridgehead group B. Thus 2-quinuclidone, or 1-azabicyclo[2.2.2]octanone-2, may be written in fractal notation as $f_{N/CH}(2)_2$ (CO1) or as $f_{CH/N}(2)_2$(1CO). In these examples it is clear from the valence of the subscript after the f that $(2)_2$ means two ethylene chains and not one tetramethylene group. If the connectors are all identical, only one need be given, so that triethylene diamine becomes simply $f_{N/N}(2)$.

Example of Fractal Notation

$(MeO)_8f_{CMe.N}(\varnothing.3)C{\equiv}C$

Peripheral group: eight MeO
Connectors: no connector, then three methylene units
Junctures: trivalent CMe, then trivalent N
Core: disubstituted acetylene
Conventional formula:

$[(MeO)_2CMe\text{-}CH_2CH_2CH_2]_2N\text{-}C{\equiv}C\text{-}N[CH_2CH_2CH_2\text{-}CMe(OMe)_2]_2$

In this example (in contrast to the one on p. 183) there are equal numbers of connectors and junctures, so the last step is from juncture to core.

Index of Molecular Fractals

One generation:

t-$Bu_6f_{C\cdot(3)}$(p-C_6H_4.m-C_6H_4)	43
t-$Bu_6f_{CLi(3)}$f(p-C_6H_4.m-C_6H_4)	43
H_2Nf_N(2NH2.2)NH	59

Three generations:

$Ph_{12}f_{P.CH.P}$(Ø.Ø.1)	39
HOf(1.NHCO.5)CH=CH	32
HOf(1.NHCO.6)C≡CH	31
HOf(1.NHCO.6C≡C)Ø	31
HOf(1.NHCO.6)C≡C	31, 33
$(MeO_2C)_{12}f_{sym\text{-}C6H3}$(Ø.TsN1.TsN1)	53
(2,6-di-t-butyl-1-oxophenyl-4-)$_8f_{13C(3).C}$(Ø.p-C_6H_4)	44
HOf(1.NHCO.2O1)C-C_5H_{11}-n	29
$(RO)_{36}$f(3.3O-4′-[(2,2′:6′2″-terpy)$_2$Ru]-4′-O3.NHCO2O1)	63
$(bpy)_{24}f_{Ru(3)}$(Ø.2,3-dpp.2,3-dpp.2,3-dpp)	87
(Me-2,3-dpp′)$_{12}f_{Ru(3)}$(Ø.2,3-dpp.2,3-dpp)	85

Cyclic structures:

$f_Ø$(ClCOCHO2O2OCHCOCl)$_2$	40
$f_Ø$(-CHGO2O2OCHG-)$_2$, G- = $RO_2Cf_{n\text{-}BuN.sym\text{-}C6H3}$ (1O2O2O2NHCO1.CO)NHCO-	40
$f_Ø$(2NH)$_6$	42
$f_Ø$(2NCO-p-C_6H_4OR)$_6$	42

REFERENCES

1. G.R. Newkome, Z.-Q. Yao, G.R. Baker, and V.K. Gupta, *J. Org. Chem.*, **50**, 2004 (1985).
2. Topterm, Inc., Columbus, OH, now provides a commercial naming service for chemicals.
3. G.D. Mendenhall, S.X. Liang, and E.H.-T. Chen, *J. Org. Chem.*, **55**, 3697 (1990).

4. An alternate naming system has appeared: see G.R. Newkome, G.R. Baker, Y.K. Young, and J.G. Traynham, *J. Polym. Sci, Part A: Polym. Chem.* **31,** 641 (1993).
5. U. Heimann and F. Vogtle, *Liebigs Ann. Chem.,* 858 (1980).
6. a. D.A. Tomalia, A.M. Naylor, and W.A. Goddard III, *Ang. Chem. Int. Ed.,* **29**, 138 (1990). b. D.A. Tomalia, *Aldrichimica Acta,* **26**, 91 (1993).
7. Z. Xu and J.S. Moore, *Ang. Chem. Int. Ed.,* **32,** 246 (1993).
8. K. Shahlai and H. Hart, *J. Am. Chem. Soc.,* **112,** 3687 (1990).
9. G.T. Visscher, D.C. Nesting, J.V. Badding, and P.A. Bianconi, Science, **260,** 1496 (1993).
10. J. Zhang and J.S. Moore, *Polym. Prepr.,* **34**(1), 120 (1993) and following papers.
11. Z. Wu, S. Lee, and J.S. Moore, *J. Am. Chem. Soc.,* **114,** 8730 (1992).

Index

www.ingramcontent.com/pod-product-compliance
Ingram Content Group UK Ltd.
Pitfield, Milton Keynes, MK11 3LW, UK
UKHW021333070726
13610UKWH00011B/60
* 9 7 8 9 4 0 1 1 0 0 0 6 9 *